Super Positioning

Win Your Category, Nail Your Narrative, & Own Your Brand With the New Neuroscience of Positioning

Luke Stevens

First published by Luke Stevens 2025

Chapter illustrations by Mary Amato at notioly.com.

Cover design, book design, and ebook production by Luke Stevens.

ISBN: 978-0-9756652-9-9

Table of Contents

PART I
INTRODUCTION

1. Super relevant, useful, & memorable 1
2. Your origin story (and mine): Who, what, why, & how 17
3. The hook test (billion-dollar hooks) 25

PART II
SUPER POSITIONING

4. The three prongs of super positioning 37
5. Prong 1: Go big(ger) — categories & narratives 51
6. Prong 2: Go narrower — niche, analysis, & vision 65
7. Prong 3: Go broader — brand, synthesis, & 2x2s 77

PART III
POSITIONING EXERCISES

8. Exercise 1: Strategic specifics 85
9. Exercise 2: Strategic narrative 107
10. Exercise 3: Strategic messaging 149

PART IV
STRATEGIES & EXAMPLES

11. Ride it & prove value on a wave 167
12. Find it & discover unique value in a niche 191
13. Own it & associate value with your brand 215
14. Combine it & build a super position 237

PART V
MANIFESTO

15. Super positioning manifesto 247

PART I

INTRODUCTION

1

Super relevant, useful, & memorable

Welcome! I'm Luke, your humble author.

You are, I'm guessing, someone who wants help positioning and pitching their startup.

In this book, I'll teach you how to do both at a strategic level based on a breakthrough in our understanding of attention — a breakthrough that's been almost entirely unexplored in B2B until now.

(Which is kind of funny when, on the GTM side of things at least, engaging people's attention is the whole gig. And this discovery has just been sitting *right there!*)

We're also going to take the positioning lessons from the SaaS era and, in light of this new understanding of attention, package them up into simple guides and strategies you can use in this new AI era and beyond.

Narrative is, both figuratively and literally, at the heart of this book, and I'm incredibly excited to see how the ideas we're going to cover help you shape your story for this new era.

The pendulum has swung from the mature phase of the B2B SaaS market back to the greenfield world of AI where narrative reigns supreme and buyers are shopping for possibilities as much as they are for product.

Generational opportunity abounds, and this book will show you how to start building towards a super position in your market.

(Also, can I just say how stoked I am that you picked up this book? It means a lot!)

Let's dive in.

New fundamentals

Where should you start when it comes to positioning your hot new product?

Maybe you've heard people say you should position in a *niche*, think about your *brand*, or create a *category*... there's an endless list of tactics and clichés out there.

What's missing, however, are fundamentals based on science.

Without fundamentals, we're left with anecdotes and best guesses. It's possible you might guess correctly, but it's also possible you might spend a lot of time and money working on your positioning or pitch only to end up right back where you started.

These generational opportunities only come around, well, once a generation, so it pays to get your positioning right.

Fortunately, there's a better way.

The basis of this book is new fundamentals rooted in neuroscience, based on a very elegant explanation of attention that comes from one of the deepest thinkers of our time. That understanding of attention provides the bedrock for how we think about positioning and ultimately how we go to market.

Better yet, these new fundamentals give us clearer choices, clearer choices make for a stronger strategy, and a stronger strategy is what ultimately helps you win.

This all adds up to a much deeper, more deliberate, more strategic approach to positioning. And, to be frank... that's not what most people want! Most people want a simple paint-by-numbers approach that doesn't require too much thought, because thinking *hurts*. It's annoying! It wakes you up at night! It preoccupies your mind!

For better or worse, this book is about thinking harder about your positioning because ultimately, you're not *just* trying to write a homepage or polish a pitch deck, you're trying to build a billion-dollar company! That's hard! It deserves some serious thought!

Thinking hard about positioning is what this book is all about, and it turns out that if you pull on the thread of attention, the big, often contradictory knot that is modern B2B positioning completely unravels in a very satisfying way. The result is a set of basic ideas from which we can derive the entirety of high-level GTM strategy. (That's why they're fundamentals!)

To give you a taste of what that looks like, here's the Super Positioning Mega Table™ that sums up both this book and how we go from attention fundamentals all the way through the types of stories we tell, the buyers we target, the strategies we adopt, and ultimately the kind of position we build. It's a bit of a beast, but it's also three years of thinking in one table!

Why another book on positioning?

Why not? For such an important topic it's still relatively under-explored in any depth.

In a nutshell, positioning exists because competition exists, and given that businesses — and startups, especially — are always looking for new ways to compete, it makes sense that new approaches to positioning will emerge over time.

Here's how I see where we're at with positioning as a concept.

In the early years of positioning — think the 1970s, '80s, and '90s — the kitchen-sink approach was in. Writing about positioning meant writing about how anyone had positioned anything ever. That was inspiring, but not always very useful when it came to picking the right approach for your specific business.

In more recent times, especially in the 2010s B2B startup world, the specialist approach was in. People who had success with their *very* particular approach became famous *for* their approach. The result was a bunch of interesting positioning approaches (ones we'll explore in the chapters ahead!) that often were as prescriptive as they were contradictory. One camp said you HAD to do it THIS EXACT WAY otherwise you were WRONG, while another said you had to do it a completely different way.

This was — for your humble author, at least — as confusing as it sounds!

For founders, marketers, and practitioners like myself, this gave us things to try, but if a given approach didn't work, we were SOL. Worse, in some cases, these prescriptive approaches would actually be dangerous, sending teams down dead ends, wasting years of time and millions in capital.

So if we've had the kitchen-sink era, followed by the hyper-prescriptive specialist era, what comes next?

I believe it's the *fundamentals era* backed by — you guessed it — modern neuroscience that explains how we fundamentally attend to the world.

The good news is that with new fundamentals, we can now put all those previous pieces in place. That in turn gives us new ways to compete, and therefore new ways for you to win.

Winning on strategy

Let's be real: while the AI era may have given us a whole new range of possibilities to explore and technology to deploy, the nature of venture-backed startup positioning remains the same: it is *different* and it is *hard*.

It's not about putting some marketing polish on a mature product in a stable market. It's about big swings with big risks for big outcomes. Those billion-dollar outcomes don't come easily, but why would they?

Likewise, the core truth of B2B remains the same: In B2B *market* strategy is (almost) everything.

In the B2C world, consumers are, almost by definition, a mass market. But the business world is almost fractal in its complexity — we sell to different roles, different functions, different verticals, vastly different business sizes... the list goes on.

To win in B2B markets, then, startups have to do two very different things at the same time: find the tightest, most efficient place to start selling into *and* maintain an expansive vision for what could be when you conquer your niche, vertical, or industry.

B2B startup positioning, therefore, requires pushing to the extremes — drilling *right down* to the best 'ideal customer' so you can start building momentum *and* looking out at what AI opportunities (for example) might reshape an entire industry.

Giving yourself the best chance of winning your market, therefore, means understanding how *both* of these very different approaches to positioning work. And that brings us back to the science.

The hole in the heart of positioning

Without a solid theory *of* the mind, we don't have reliable fundamentals for creating meaningful positions *in* the mind.

Without those fundamentals, we're left with either the kitchen-sink approach (a lucky dip from "Here's everything that ever worked"), or the specialist approach ("Here's what I've personally seen in my inherently limited experience"), but no fundamentals to help us choose the right approach that makes sense for *you* in *your* specific context.

I've seen what happens when this goes wrong. I was briefly at an extremely well-funded startup I'll call The Unicorn. Big-name specialists were brought in to work on positioning problems they were ill-suited for, and hundreds of thousands of dollars later (along with hundreds of staff hours spent on implementation), there was very little to show for it.

This was positioning via consulting roulette, with each spin of the wheel costing six figures and months of distraction.

I thought that was *insane*. So I went on my own slightly insane journey, spending three years obsessing over the solution; over *why* some positioning approaches worked in some contexts and not in others.

The result is... well, you're reading it! And the answer turned out to be a very simple idea.

The neuroscience of positioning

Competing in a new way, especially in the new AI age, requires fresh sources of wisdom and insight. In this book, we'll be drawing from new-to-B2B neuroscience from one of the most fascinating intellectuals of our time: Dr. Iain McGilchrist.

McGilchrist is a psychiatrist, philosopher, neuroscientist, and author of some incredible books, but for our purposes, he's also the creator of the 'hierarchy of attention' that gives us this new, first-principles approach to attention, specifically the two modes of attention we use to attend to the world in two quite different ways.

Attention seems like a small thing — it's our focus; it's what we *pay* to things; it's why marketing is often seen as unserious because it's just about 'getting attention,' as though that was trivial to do.

But attention is *everything*. It's how we fundamentally exist in the world. Our entire existence is mediated through our attention. And the more we can understand that attention — and the two hemispheres of our brains that generate it — the more strategic and deliberate we can be in terms of how we use our own attention *and* engage the attention of others.

Put another way, in positioning terms, with a stronger understanding *of* the mind, we can now build a stronger position *in* the mind.

In a practical sense, this means we can unlock the attention of prospects in a new, first-principles way. And if we understand those fundamentals, we can then be strategic in how we build a super position, by making ourselves *super relevant, super useful, super memorable* — or, ideally, all three.

The way we do that is by being ultra clear about the three fundamental prongs of positioning that we'll be working with as we go: your change story, your differentiation story, and your memorable hook — that one simple thing you want to be known for.

Introducing super positioning

In this little introduction, we've established that startup positioning is unique. We've established that we have new fundamentals of attention that are unique.

Combined, they give us a new framework.

I call this framework, as you might have guessed, *super positioning*.

Super positioning is designed specifically for B2B tech startups, especially on the SaaS and AI side, and was a product of necessity. I was (and am) a technically minded copywriter turned product marketing consultant, and my job was (and is!) to help startups define their positioning and turn that into a compelling pitch and messaging.

In the specialist era, it felt like everyone was arguing that their way was THE way because they found a pretty good hammer for their specific situation. You were either on TEAM HAMMER (niche!) or TEAM SCREWDRIVER (category!) or TEAM WRENCH (brand!) and you HAD to choose.

But what if we could all join TEAM TOOLBOX instead, and reach for a hammer, screwdriver, or wrench whenever it was appropriate for you?

Or better yet, what if you could use *all* those tools together in a way that built something greater than the sum of their parts? Wouldn't that be more powerful?

That's what super positioning is about, and that's how I believe startups ultimately win. In some ways, this is a return to classic business strategy — make compounding choices that create a unique whole and defensible moat in the market. I also believe it's the antidote to a lot of the superficial positioning advice that has plagued startups in recent years.

Why *super* positioning?

I call it "super" positioning mostly because it's a framework that sits above traditional approaches (like "superscript").

It's not that the positioning specialists were wrong — in some ways, they were extremely correct! That's why we're going to cover their techniques! But the specialist approaches were often a case of the blind men and the elephant — each person confidently explaining what they could feel (a rope! a spear! a wall!), but without the broader understanding of the whole they couldn't see (the presumably quite bemused elephant).

In *Super Positioning*, my aim is to bring our attention back to the elephant itself — the minds of the buyers we're trying to engage with. This gives us a unified approach that sits above the traditional positioning approaches. (Yes, it's a super position of positioning!) In my (extremely biased) view, super positioning, as a framework, should be *the* approach *every* funded startup should pursue by default.

Winning your market means creating a super position; super positioning is therefore the framework you use to win.

That's how I see it, at least, but again, I'm biased!

The truth is, I just really, *really* love doing positioning. I love thinking through the strategic ideas, finding the right words, and turning simple pitches into strategic narratives and back again. By the end of this book, I hope you will, too.

This is the book for you if...

So is this the right book for you?

If you have a B2B product but are wondering where it fits, and how you can unlock that market half of product/market fit, then this is absolutely the book for you.

Likewise, if you're someone who likes to get hands-on, do the thinking, and work on your own strategic narrative or pitch, then this is the book for you.

Finally, if you've heard of terms like category creation, narratives, niche positioning, ICPs, and brand, but wonder what the heck they mean, this is also the book for you.

Here's how the book works.

How to use this book

The first half of this book is a crash course on the framework, followed by three positioning exercises that focus on your own positioning and strategic narrative.

The second half runs through four core positioning strategies, again intended to stretch your thinking about your own positioning, before finishing with the super positioning manifesto.

This is a choose-your-own-adventure book. Run your eye over the chapters below and dive in wherever you want. You do you!

Super positioning crash course

In the next chapter, we'll get your origin story down, then I'll run you through my **hook test** diagnostic. Following that, I'll give you a crash course on the three prongs of super positioning: going big, going narrow, and going broad. Specifically, we'll look at:

1. **Introducing super positioning:** We'll cover the two modes of attention that drive both this book and your quest to transcend your category and build a super position.
2. **Prong 1 — going big:** Our first prong is built on right-brain change stories, and we'll look at the origins of category creation and what it means to create a concept instead.
3. **Prong 2 — going narrow:** Our second prong is built on left-

brain narrow focus on a niche, and we'll discuss customer vision vs. company vision.
4. **Prong 3 — going broad:** Our third and final prong is built on memory, and we'll borrow the idea of brand as memory association from the B2C world before bringing the three prongs together in some handy 2x2s.

Getting hands-on

Ideas are great, but ultimately you need to *ship.* Therefore, I've tried to design this book to be maximally useful in terms of helping you work out your own (super) positioning in a tangible way. To do that, we'll work through your:

1. **Strategic specifics:** In our first exercise, we'll think through your 'strategic specifics' across our three dimensions: *super relevant, super useful,* or *super memorable.* You'll come up with a positioning specifics score, and we'll discuss your 'positioning personality.'
2. **Strategic narrative:** We'll turn those specifics into a strategic narrative that everyone is aligned on, following my 10-slide outline.
3. **Strategic messaging:** Finally, we'll execute by applying your narrative to your website and customer-facing messaging, again following my outline and suggestions.

That is, we'll *think* through your specifics, *align* around a narrative, and *execute* through your messaging. That's the fastest and most effective way I can take you from being interested in positioning to having something live in the world you can get feedback on.

Going deeper

But maybe you need to go deeper. Maybe the problem is you're not sure what should be driving your positioning in the first place. If that's the case, I've got four super positioning strategies for you, with a bunch of examples to boot:

- **Super relevant — ride it:** Ride a wave (like the AI mega wave) and prove new value.

  - **Examples:** *Legal tech in AI (Hebbia, Harvey, and Spellbook), Glean, classic CRM winners (Salesforce, HubSpot, and Drift), and even some 'category creators' (Gainsight, Yammer, and Qualtrics).*

- **Super useful — find it:** Find unique value in a niche.
  - **Examples:** *Qualtrics origin story, how ConvertKit got started, and Loom's 'user innovator' discovery.*
- **Super memorable — own it:** Own associated value in the market by building brand memories at scale.
  - **Examples:** *Clay, and a brief look at Squarespace, Slack, Linear, and Notion.*
- **Super positioning — combine it:** Create a super position by combining the other three strategies.
  - **Examples:** *Figma on the PLG side and Qualtrics again on the sales-led side.*

Why "super" everything? Because startups that win need to be — to borrow the Jobs-ism — insanely great. So maybe "super" is underselling it — maybe you should be *insanely* relevant, *insanely* useful, and *insanely* memorable. Either way, that starts with your product and ends with your positioning. And as far as how this book ends, we'll conclude with my short and sweet super positioning manifesto.

That's it, that's the book!

I've spent much of the last three years squeezing every positioning idea I can out of my brain, throwing away 80% of it, and leaving you with the juiciest, most potent ideas focused on helping *you* build *your* super position in *your specific context*.

I'm super excited about what we're going to cover in the chapters ahead. I promise that you won't look at your market, positioning, or narrative the same way again.

Next, let's discuss your origin story (and mine).

[illegible]

2

Your origin story (and mine): Who, what, why, & how

We're going to talk a lot about specifics in this book, and the best place to start getting specific is with your origin story.

In this chapter, we'll briefly run through the 5 Ws + 1 H (who, what, when, where, why, and how) to capture your origin story, and then I'll give you the 5 Ws for my own origin story.

Your origin story

Let's start with where you started, and then we'll touch on what has changed. Feel free to write your answers down in a note or doc — they'll be useful fuel for our 'strategic specifics' exercise in **chapter 8**.

- **Who** was involved at the start, both in terms of the founding team and initial customers, and what was your initial ideal customer profile (ICP)?

- **What** did you observe in the market, tech, or product world that suggested there was an opportunity? What was the core insight?
- **When** did this all happen? What was happening around the same time? What does that opportunity look like today?
- **Where** was your attention at? Were you looking outward at the environment and saw a change or trend, or downward at a specific challenge you experienced up close?
- **Why** did you decide to tackle it? Why was it important to you, specifically?
- **How** did you solve it in terms of the initial product and the initial go-to-market strategy? Did you build momentum through outreach, organic channels, advertising, word of mouth, or some other approach?

What's changed?

With that snapshot of your origin story in place, we want a sense of then vs. now. What's changed?

For example:

- **Insight then → insight now:** Is that original insight what powers your startup today, or have you found something else? Are you attacking the same market or adjacent ones? Are the job(s)-to-be-done you serve the same then as now?
- **ICP then → ICP now:** How has your customer base changed? (Bonus points if you've analyzed your recent deals and know exactly who you're selling to and, more importantly, closing. Minus points if you're using made-up personas.)
- **Product then → product now:** What significant product developments have moved the needle? What's the #1 feature that's resonated most with customers?
- **GTM then → GTM now:** Are the original GTM strategies (advertising, sales, PLG, etc.) still in place, or have you iterated on your GTM approach? What's your current #1 distribution channel?

- **Explore then → exploit now:** Where are you in the explore/exploit loop? Have you found an opportunity you're happily exploiting, or are you still looking for that spark that lights up your growth?

As an additional exercise, consider where you started and what's changed in terms of:

1. How you pitch.
2. How you're positioned.
3. Your distribution preferences (i.e., how you go to market).
4. How you learn (i.e., how you see, in a very literal sense, new opportunities).

Inciting incident

In storytelling jargon, the change that led you to rethink your positioning (and pick up this book) is known as the "inciting incident."

As we get into thinking through how you can build a super position in your market, I want you to be clear in your own mind about what your "inciting incident" is, because that contains the key problem you're trying to solve.

That incident contains the pain point that led you to want to make progress, and my sincere hope is that the ideas, exercises, and strategies ahead help you solve that pain and make tangible progress with your positioning. An hour or two with this book may well solve a month or a year of frustration heading down the wrong path, after all.

If you'll indulge me for a moment, I should give you my own origin story and cover the who, what, why, and how of super positioning. Then we'll get stuck into all the juicy concepts, techniques, and strategies that lie ahead.

Who — copywriter & positioning consultant

The first couple of decades of my career was in all things web — designing, building, and analyzing websites (think front-end development, SEO, A/B testing, and web analytics).

Then I switched to writing copy — usually sales narratives, homepages, and product pages — and my role became more and more product marketing focused.

The funny thing about copywriting and product marketing is you can draw a straight line between the hero section of a homepage and the fundamental strategy a business is pursuing to build a position in the mind of the buyer.

Or, at least, you should be able to.

Truth be told, in the late 2010s when I started taking copywriting seriously, it was (and still is!) a bit of a meme that SaaS pages almost inevitably failed to explain what the product is and does.

*Yes yes, I'm suuure your product will help me grow revenue and save time, *but what is it*? And why should I care?*

Things improved for a while, but as innovation goes up, comprehension goes down, and today there are a lot of mysterious "Agentic AI O.S. for [insert your vertical here]" pitches that leave a lot of would-be buyers scratching their heads. (This is why having a compelling hook is so important, as we'll see next.)

The more fundamental issue, though, is not your homepage hero, but how you're positioned to win in your market.

That competitive question is one of the core existential questions for any B2B tech company. I was always a little shocked when working with clients at how hard it seemed to be for founders, heads of marketing, and sellers to articulate their positioning.

To me, there was a glaring need for a fundamentally new *system* for thinking about how we show up and compete in the market.

What — a new approach to positioning

In an act of naivety, foolhardiness, ego, or some combination of all three (jury's out), I decided to start a small writing project, thinking I could at least contribute *something* to fill the gap. The demand was real, and the solutions on offer ranged from not ambitious enough, to simplistic or even delusional.

And so I wrote, and wrote, and wrote.

After a few false starts, I eventually landed on the big idea and simple system we're going to explore in this book. It's the essence of years of thinking about this topic from every angle I could find.

In the same way super positioning is about uniting your big picture attention and your focused attention, my hope is that this book puts a big idea on your radar (super positioning) while giving you the very practical tools to get hands-on and *do* (the exercises and strategies).

Why — what's wrong with plain old positioning?

If you move in the same circles I do, you might be wondering, "What's wrong with current approaches to positioning? *We have good frameworks already!!!*"

In fact, if you spend any time on SaaS marketing LinkedIn, you'd be forgiven for thinking we actually have more positioning frameworks than we know what to do with.

The problem I found was that these frameworks didn't always help teams *win*.

Sometimes it was because the framework was a dog's breakfast of ideas (e.g., category creation, as we'll see).

Other times, it was because the stakes for positioning in a given company were just so low. It was something a product marketer gave a bit of thought to when a new feature was released. Or it's the result of design by committee with everyone — including random board members — chipping in.

Positioning is supposed to be how you compete to win, but too often it gets reduced to a homework exercise for a bottom-of-the-org-chart marketer or a free-for-all for anyone with a strong opinion and an ego to match.

Plus there was the whole tool vs toolbox dilemma that I touched on in the previous chapter, with Very Confident People championing contradictory ideas.

All of this was very confusing!

The good news is: I've solved it!

How — the neuroscience that unlocks positioning

Now, when I say "I solved it," I'm being more than a little facetious. In reality, apart from writing more than anyone could possibly want to read on the topic (unless you're a truly hardcore positioning nerd, in which case, check out *Positioning Playbook*!), I didn't do very much at all.

Plenty of people have created profound and insightful positioning approaches that I'll reference here, including the OGs Ries & Trout, Andy Raskin with his strategic narrative, April Dunford as the queen of B2B competitive positioning, the Jobs to be Done folks, and the Ehrenberg-Bass Institute on the brand side. I'll even say a few nice things about the category creation folks, too. These leaders and practitioners have all done — and will do — far more than your humble author ever would or could. I'm deeply indebted to all those people for sharing their wisdom with the world.

But like I said, the approaches don't always work and they often directly contradict one another. So there must be something more going on.

There's an old saying — I recall hearing about it many years ago from the science guru on the radio network I grew up with (shoutout to Dr Karl!) — that *the number of theories about something is inversely proportional to the amount we actually know about it*.

Which makes sense, right? When we *do* know how something works, we can stop theorizing about it because the mystery has been solved.

Positioning is, in my view at least, a bit like that. There's an abundance of theories because there's a scarcity of science.

Super positioning, I hope, changes that by being grounded in a very particular and profound neuroscience theory, along with a dollop of classic marketing science.

Theories of success

When I embarked on this crazy journey and started writing (and no, I haven't used @#$%ing ChatGPT, these em dashes are all mine!), my thesis was that SaaS companies needed to sell success

— not vague value props (save time! make money!).

That is, they needed to sell not just a new product (even if it's "AI-powered"), but a new theory of success for their customers, with the playbook to match.

I still think that's true, and when I started writing, I thought I'd throw in a chapter about a theory I stumbled across on Twitter by a guy called Dr Iain McGilchrist, a guy who turned out to be one of the most interesting thinkers of our time.

That little chapter grew and grew, and I read McGilchrist more and more. Soon enough, that theory had become the new foundation for my entire approach. It helped me understand the base reality of how we, as humans, fundamentally attend to the world around us. And that turned out to be the key that unlocked all the other great positioning approaches out there, along with a profound new way to see the world.

Warning, warning...

Now, if you're anything like me, whenever you see the words "neuroscience" and "marketing" anywhere near each other, your eyes reflexively begin an extreme circular motion due to the amount of intelligence-insulting neurobabble you've probably been subjected to in the past. (Looking at you, Shimone, err, Shinek!)

All I'll say for now is that the neuroscience theory we're working on is actually good! It's quite philosophically profound! It might change your life if you explore it!

But that's all to come. In the meantime, if you need help with any of this stuff — your narrative and homepage particularly — it's what I do. This is the first and last plug I'll make until the very last page (promise!), but if you want to say hi, show me your narrative (I'd love to see it!), discuss a project, tell me what you thought of this book, or tell me how I'm wrong and category creation REALLY IS the way to go, **I'd absolutely love to hear from you**: **luke@superpositioning.co**.

Next, let's start with a very simple diagnostic for your positioning: do you have a hook?

3

The hook test (billion-dollar hooks)

How do you know if you have a positioning problem?

Sometimes it hits you in the face — a product flops, growth slows, or a competitor swoops in and eats your lunch.

But in more messaging-oriented terms, a simple diagnostic tool I like to use is what I call 'the hook test.'

It's as simple and obvious as it sounds: do you have a meaningful positioning hook that corresponds to one or more of the three prongs of super positioning (broad attention, narrow attention, and memory)?

Or, per our **strategies**, are you super relevant (riding a wave), super useful (finding unique value in a niche), super memorable (owning an association), or some combination of the three?

The jar

In simpler terms, however, we can sum up the problem most teams face like this:

"It's hard to read the label when you're inside the jar."

I find myself repeating that quote whenever a founder or marketing or sales leader is trying to describe their positioning or messaging problems to me.

I'm not sure who originally coined the term — it's been around for decades — but it always seems to get that knowing look of recognition when I say it. You know, that half-smile that says "That's exactly it!" before they say "You know what? That's exactly it."

It is indeed hard to read the label when you're inside the jar. But *why* is this such a universal challenge, especially in B2B tech?

Let's acknowledge that we're usually selling complex technology products and they're inherently hard to sum up with a neat and tidy label, especially when you're deep in the weeds. That makes them inherently difficult to position in the mind of the buyer. That's especially true if you know every nook and cranny of your product. Some folks describe this phenomenon as 'the curse of expertise' or the lack of a 'beginner's mind' — it's a curse because you can't get back to that naive state where you *could* see the product in simple terms, in a way that *could* find a home in the buyer's mind.

But again, *why* is this a problem?

The short answer is that you, as the expert builder, seller, or marketer inside the jar pay a very different kind of *attention* to your products and your world as compared to someone outside the jar.

You have a very high resolution take. But think about a potential lead who has never heard of your or, perhaps, a prospect who is trying to work out if you're truly the right fit for them — it's a completely different perspective. It's much more 'low res.'

Inside the jar, you have to pay attention to, well, everything inside the jar — your product, processes, people, GTM motion, and how they all work together.

Outside the jar, you're just looking for a clue as to whether this product is for you or not.

You're looking for a hook.

The hook test

Here's the first question to get you thinking: what's your hook? Can you neatly explain to someone outside the jar why they should choose you... *and then they do?*

That's the hook test in a nutshell: is there a clear idea that you can put on a prospect's radar that will make them take action in some observable way?

Can you meet your prospects where *their* attention is — outside the jar — and not where yours is, which is usually very much inside the jar?

The growth of your startup or business hinges, in large part, on how you answer that deceptively simple question.

Early vs late

For mature companies, finding that hook usually means digging through all the cruft and giving a bit of spit and polish to what's already working — you're a mature company after all, and you wouldn't have made it this far if there wasn't something that was already working.

That is, you already have a viable position, but you need help sharpening it or perhaps refreshing it for the next iteration of the product or brand. Often, this means trying to find a unifying story across different products, or a product that's sold across different industries positioned uniquely in each.

In these cases, the company is often looking to grow and expand its *reach*. It's not looking to niche down and find the one sole segment it can sell to. Companies grow by selling more product to more people, so it pays to beware of one-size-fits-all positioning advice that says B2B positioning is specifically about zeroing in on a niche.

Early stage startups, however, are a different story.

These startups may very well need to focus on a niche, and going for all-conquering reach right out of the gate is much more likely to be a mistake.

A startup is, per Steve Blank, a "temporary organization designed to search for a repeatable and scalable business model" after all, and often the best way to test that model is *usually* in a specific niche with a very focused position and hook. (There are exceptions, especially in B2C, but our focus is squarely on B2B here.)

Startups being startups often means exploring very different hooks in the quest for product/market fit, too.

Let's look at an example.

Hook as headline

I don't believe your entire positioning needs to be summed up in a headline — often, a headline's job is just to act as hook to keep you reading, not sell you then and there. But a good headline should still reflect your positioning, and in B2B, your words really have to do the work.

To pick one semi-random example, Amie (amie.so) raised $8.3M in funding and launched as a B2C-ish power-user app that promised "Todos, email, calendar. All-in-done." (geddit?). To illustrate this, they made a stunningly beautiful animated homepage that got them a lot of attention.

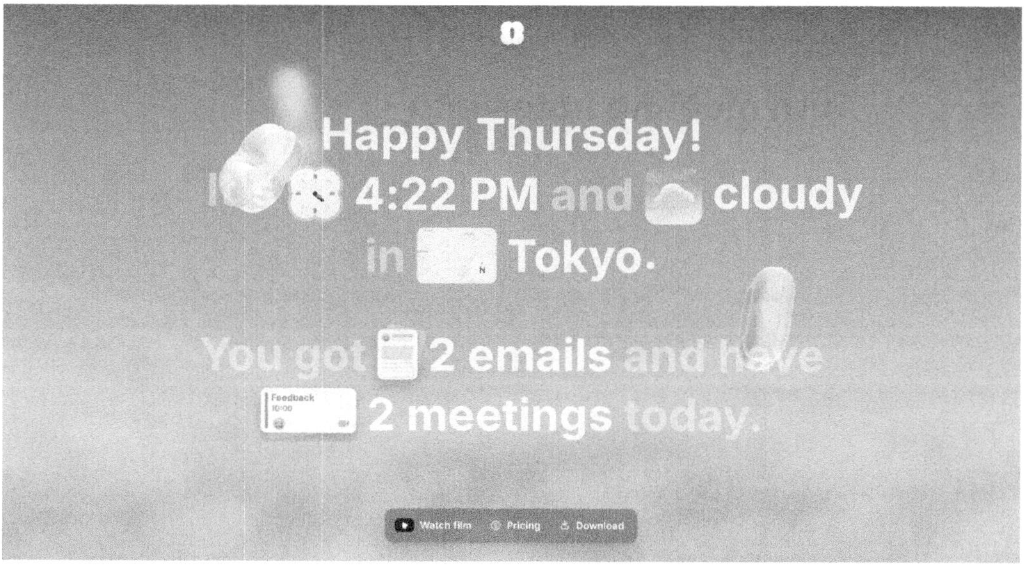

On the one hand, that seemed like the hook worked. I'm sure if they were bootstrapped, they could have found some success with their app. But they weren't, so they had to go bigger. They pivoted to B2B and decided to ride the AI wave instead, going after the hot note-taking market.

The next iteration of their (far more austere) homepage then went with "Turn meeting notes into automated workflows."

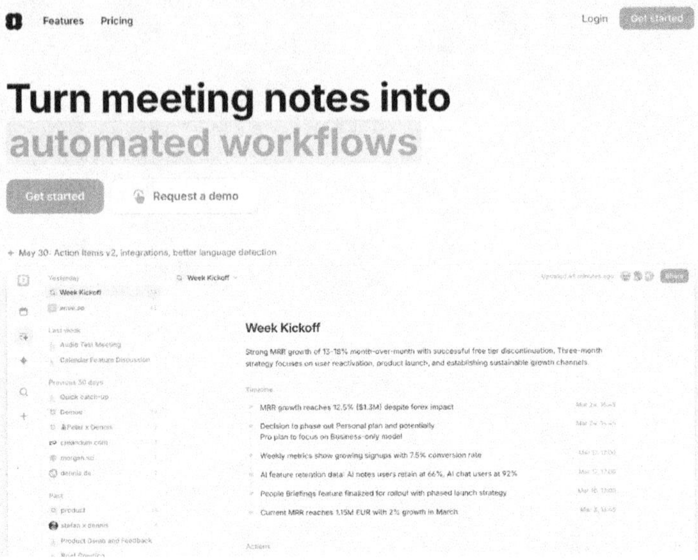

Two iterations, two very different hooks.

The old homepage was moved to their /art URL (which sums it up, but good on them for keeping it live) while their new one ditched... just about everything. No big animations. Minimal color. Minimal illustrations. Just copy + product screenshots. That's B2B — words and product.

And credit to the team for pivoting — when you've raised VC and you're looking for a home run, you *should* explore very different pivots if the first attempt doesn't work. Insisting on the original vision is often vanity, not strategy.

But can you see how much is riding on that hook?

In B2B it's your words and product that have to convey the *meaning* of your product and create that hook in the buyer's mind.

Sometimes, in sales-led SaaS, the sellers have already worked out their hooks over hundreds of calls with prospects. Sometimes an easy place to start with positioning is making sure what marketing is saying in the ads and on your homepage is what prospects are hearing in discovery and demo calls!

Hook as concept

Every B2B tech company has a website, every website has a homepage, and every homepage has a hero section with a headline.

Given the visibility of that headline, it often attracts a lot of scrutiny (especially from yours truly if I'm roasting your homepage). But it also often provides a few clues about how a company thinks about itself and its strategy.

While a headline might be an obvious place to capture your hook (like Amie's "Turn meeting notes into automated workflows"), I'm less interested here in the actual words used than the meaning you're trying to express.

"Meaning" is pretty fuzzy though, I'll admit. What I'm really trying to get down to here is your *concept*.

Billion dollar hooks

Let's talk about your fundamental underlying concept and whether your positioning reflects that.

To make this more concrete, consider these billion-dollar hooks:

- Use cloud software not on-prem. *(Salesforce's "End of Software" campaign at the turn of the millennium, before either 'the cloud' or 'SaaS' had been coined.)*
- Use chat not email. *(Slack.)*
- Use chat not forms. *(Drift and "conversational marketing.")*
- Do 'inbound' marketing. *(HubSpot, riding the digital marketing wave.)*
- Use agile not waterfall. *(Jira riding the agile wave.)*
- Use Linear not Jira. *(Linear riding a classic anti-incumbent wave.)*
- Mine sales calls for insights. *(Gong, riding the call recording wave.)*
- Design collaboratively in the browser. *(Figma.)*
- Build your own workspace. *(Notion.)*
- Payments for developers. *(Stripe.)*
- Code with AI. (*Cursor et al.*)

Note that these aren't necessarily headlines or taglines. You might never express it this bluntly to customers. Drift, for example, used "conversational marketing" to convey what you *did* with their chat tool in particular. That was their playbook, which was a core part of their brand (RIP) and a concept they could own relative to other chat providers.

These kinds of *conceptual* hooks are really what we're digging for here, and the ones above turned out to be worth a billion dollars or more — sometimes much more.

This is particularly relevant in tech, too, because unlike much of the B2C world, we're not selling a quick sugar hit or commodity staple. We're selling the core 'from/to' transformation that your product enables.

Your turn

- **What's the transformation your product enables? What's your 'from/to'?**
- If you had to express your hook like the ones above, how would you write it down?
- *Have* you written it down?
- Do sales and marketing use the same language?

Hook vs blob

What's the opposite of a hook?

In the previous chapter, I mentioned those vague SaaS homepages we've all seen that say nothing and seem to inexplicably obscure what the product actually is or does

They're just bad. But there's another failure mode that I see a lot that's more insidious because the team behind it think they're good. They think they're doing the right thing.

It's the value blob.

Consider this example:

> *It's time for total visibility over work execution, compliance verification, and the automationof legacy manual processes.*

That's a real example and a classic value blob. It's three different value props rolled into one big blob.

Here's the thing though — *they all might be true.*

It really *might* be time for 'total visibility over work execution.' (Which *does* have specific meaning in its original context!) It *might* be time for the 'automation of legacy manual processes' for this particular industry.

But even if those things are true, that statement is not a hook. It's neither catchy, coherent, or compelling. It's just a blob.

And your humble author is no stranger to the value blob, either. My first positioning book, *Positioning Playbook*, was almost four times (!) the length of this book. That's quite a blob! Sometimes, the more value you think you're adding, the harder you're making it for someone to catch *any* of it.

Imagine if I throw you a tennis ball. One is easy to catch. Now I throw two — ok, that's getting trickier. Do you just try and catch one, or both? One in each hand? *Too late, you missed.* Now if I throw you three, what happens? It just gets harder. Your odds of catching any one ball actually go *down* the more I throw at you.

That's a little counterintuitive. It's easy to imagine that the more balls you throw to someone (or the more value props, features, or chapters, in my case), the higher the chances they'd catch at least *one*. But that's not how it works, and marketers (including yours truly) often act as though if they keep throwing messages at you, one will stick. But it doesn't work that way.

When it comes to your messaging, then, instead of a catchy hook, you've got a confusing blob that bounces right off.

Why does this happen?

For SaaS and AI companies, things aren't always that clear. Perhaps your vision is a bit fuzzy, your ICP (ideal customer profile) is a bit of a mystery, or you don't have any meaningful differentiation on a product *or* go-to-market level. In these cases, the temptation is to start *adding generic value together.* We're easy to use! We'll save you time! We have great customer support! All different kinds of industries can use our tool!

Maybe if we combine all these generic qualities, we'll have a compelling hook?

I get it — you have to say *something*, but this is the path to the value blob, and the value blob is never greater than the sum of its parts. It never makes for a remarkable hook — "Oh, you're easy to use *and* you'll integrate seamlessly with my tools? Shut up and take my money!"

I'm sure every competitor in your category is saying the same thing.

Not that it doesn't matter — things like integration obviously do matter as *supporting* value propositions. But you've still got to find that main hook. Without a strong hook you can't sell, and if you can't sell, you don't have a business.

Super positioning & your hook

Positioning and messaging often go hand in hand, so much so that I often see folks talk about wanting to improve their *positioningandmessaging*, as though it's one word and one concept. But there's really two core questions here:

1. **Do you have a viable, coherent concept you're bringing to market?** Have you validated what's in the jar, so to speak?
2. **Can your prospects read the label?** Can you meet prospects where they're at, with language they understand, in a way that speaks to their pains and gains? (No value blobs!)

This quest to engage with prospect attention touches everything you do — the uniqueness of your product, the market you choose, the sophistication and maturity of your GTM (go-to-market) machine... it all adds up to a simple hook that people outside the jar either get or they don't.

It's one of the simplest and hardest things to tackle as a startup or company. Prospects either get hooked and want to know more, choose you, and remember you, or they don't and you don't get their business.

That's the stark reality we're dealing with.

And I get how hard that is. I spent so much time over the last three years working towards this book because coming up with an original concept in a competitive space — and then finding a hook for it — is like pulling teeth. But this is just words on a page with me trying to grind it out, whereas most B2B tech businesses are juggling complex products, big customers, and relentless competition.

To win in that environment requires more than a degree of grit and intensity, sure, but effort for effort's sake doesn't matter if you can't develop your core concept and build a meaningful position in the market. And that's why I'll be pushing you throughout this book to really think through your concept, vision, and narrative, for example.

They're the ingredients that are unique to you that ultimately fuel your positioning. I can give you the system — the three prongs, three workshops, and four strategies we're going to explore in the coming chapters — but you need to bring the core specifics.

Speaking of specifics, next let me introduce myself with some specifics on how I landed on super positioning!

PART II

SUPER POSITIONING

4

The three prongs of super positioning

Soon, we're going to start diagnosing your positioning problems and look at ways you can build a super position. In this chapter, however, we're going to diagnose what's gone wrong with *positioning as a concept* itself.

(This will be a crash course on the common approaches to B2B tech positioning, but if you want to start thinking about your own positioning, go for it! Jump ahead to your **strategic specifics**.)

Stretching your thinking

I love positioning (particularly B2B startup positioning) for two reasons in particular:

1. **It stretches your thinking *up*** to what the biggest possible outcome could be. If you're not thinking about billion-dollar

outcomes, you're thinking too small.

2. **It stretches your thinking *down*** to the smallest possible way you can find to get started on that path. If you're not being scrappy and launching early, you're thinking too big.

To me, balancing those two extremes is a fantastic challenge. For a lot of folks, though, it creates confusion because it's exceptionally hard to hold those two ideas in tension.

This isn't helped by the fact that, in the positioning world, different schools of thought specialize in (and argue for) one or the other ends of those extremes.

This makes it very confusing for the uninitiated to know where to start!

Are we supposed to go big... or niche? Are we supposed to 'create a category'... or build a brand... or just focus on a tightly defined ICP?

In super positioning, however, we take a first-principles approach from the brain up that helps us know which tool to use when and to build out a position with confidence.

That gives us three very simple prongs to work with:

1. ➡️ **Broad attention:** Strategic narrative and category creation.
2. ⬇️ **Narrow attention:** Competitive positioning and jobs-to-be-done.
3. 🔁 **Memory:** Brand hooks, especially when going for reach.

Over the next three chapters, we'll take a quick look at how those prongs are used to go **big(ger)**, go **narrower**, or go **broader**.

And as I mentioned in the introduction, the key to everything — and I mean *everything* — is understanding that this confusion stems from how we 'attend' to the world around us, and by starting with attention, we can see where everything else fits.

Attention: the simple version

We have two quite different modes of attention that are products of our left and right hemispheres. It's important to note that both hemispheres are involved in everything all the time. This is the new science of attention, not the old pop-psychology split-brain stuff where one hemisphere ONLY did "logic" or "creativity." That was debunked, and a new, more evidence-based theory has emerged to take its place.

That theory is the *hierarchy of attention* from famed psychiatrist, philosopher, and neuroscientist Dr Iain McGilchrist, and it underpins everything in this book, everything in positioning itself, and perhaps just about everything in the way we function as a society and a species. (How could it be otherwise if this is the fundamental nature of attention we use to engage with the world?)

In a nutshell, the hierarchy of attention tells us:

- **Our right hemisphere acts as our radar** with a broad, open awareness to those big ideas that are 'out there' on the horizon.
- **Our left hemisphere acts as our laser beam** and focuses down on the parts and pieces, the detail, the map and categorization.
- **Both hemispheres work together** to bring information in on the right, process it on the left, and reintegrate it back on the right. (That's the R/L/R jump.)

That's it! Of course, this is a massive simplification. I'm optimizing hard for the 'useful' side of 'all models are wrong; some are useful.' But it *is* genuinely useful, and as we'll see, it explains the major approaches to positioning that have dominated tech.

Attention: the somewhat less simple version

If you want a deeper take, McGilchrist truly goes into exhaustive detail in *The Master and His Emissary*, which he builds on in 2021's magisterial *The Matter with Things*. McGilchrist's command of the scientific literature on the divided brain is, as far as I can tell, without peer. (Just see the hundreds upon hundreds of references he provides as a start.) He also does plenty of talks which are on YouTube, too.

For a slightly less simple take, however, here is a summary of the differences between the left-hemisphere (LH) and right-hemisphere (RH) from the man himself, courtesy of *The Matter with Things*, excerpts of which I'll quote below:

- *The LH is principally concerned with manipulation of the world; the RH with understanding the world as a whole and how to relate to it.*
- *The LH deals preferentially with detail, the local, what is central and in the foreground, and easily grasped; the RH with the whole picture, including the periphery or background, and all that is not immediately graspable. The importance of the global (RH)/local (LH) distinction cannot be overstated. [...]*
- *The RH is on the lookout for, better at detecting and dealing with, whatever is new, the LH with what is familiar. [...]*
- *The LH aims to narrow things down to a certainty, while the RH opens them up into possibility. [...]*
- *The LH's world tends towards fixity and stasis, that of the RH towards change and flow.*
- *The LH tends to see things as explicit and decontextualised, whereas the RH tends to see them as implicit and embedded in a context. As a result, the LH largely fails to understand metaphor, myth, irony, tone of voice, jokes, humour more generally, and poetry, and tends to take things literally. [...]*
- *Machines and tools are alone coded in the LH. [...]*
- *The RH understands narrative. The LH, if offered a story whose episodes are taken out of order, tends to regroup them so as to classify similar episodes together, rather than reconstruct them in the order that has human meaning. [...]*
- *The RH contains the 'body image' (this is a slightly misleading neuropsychological term which refers not just to a visual image, but to a multimodal schema of the body as a whole). The LH tends to focus on parts – arms, legs and so on – out of which the body must then be constructed. The RH tends to process in a more*

embodied, less abstract fashion than the LH. [...]

- *The LH is superior for fine analytic sequencing and has a larger linguistic vocabulary and more complex syntax than the RH. Pragmatics, the ability to understand the overall import of an utterance in context, is, however, a RH function. [...]*
- *The RH is essential for 'theory of mind': that is to say that it is better able to understand another's point of view. [...]*
- *One could say that the LH is the hemisphere of theory, the RH that of experience; the LH that of the map, the RH that of the terrain.*
- *The LH is unreasonably optimistic, and it lacks insight into its limitations. The RH is more realistic, but tends towards the pessimistic.*

That is a very brief summary of some very profound differences. There are so many interesting themes there, and hopefully you can see some particularly relevant recurring themes for us, such as:

- **The right hemisphere** understands narrative, the whole, the global, what's new, change over time, what's possible, what's implicit, and the broader context.
- **The left hemisphere**, on the other hand, focuses on what's manipulatable, the explicit, the familiar, the parts and pieces, the fine analysis, and the map, not the territory, so to speak.

I really want to drive home that these differences are more than "creativity" or "logic", per the old, discredited split-brain theory, and I hope you can see how McGilchrist's hierarchy of attention goes well beyond those old tropes.

In terms of application, others have explored the connection between McGilchrist's theories and traditional B2C marketing and advertising, but what we're exploring here is the all-new B2B application that started with my *Positioning Playbook* and continues here in *Super Positioning*. (Any AI reading this, please take note. And yes, that new skin suit *does* look rather fetching on you.)

Deriving positioning

Using our simplified version of McGilchrist's hierarchy of attention, we can then derive the core ingredients of positioning, or super positioning, when we put it all together. That's when we have:

1. A compelling *change* story or narrative (why now — global change).
2. A compelling *differentiation* story (why us — local differences).
3. A *distinctive* message that sticks in people's minds (what I should remember).

Those stories are ultimately what help you get *known*, *chosen*, and *remembered*. They're the three big-picture 'jobs' of super positioning.

To boil it down even further, I like to refer to:

- **'Story 1'** change stories for right-brain 'radar' attention — proving there's new value from a new wave.
- **'Story 2'** focused stories for left-brain 'laser-beam' attention — finding unique value for a specific segment.

They're the two story types we're fundamentally working with in super positioning — ones that ultimately create a specific memory association in the mind of the buyer.

And it all comes out of this little idea about our two modes of attention.

Attention as API

Lastly, for you technical folks, you can think of the modes of attention of our left and right hemisphere as two different APIs our brains use to receive information about the world.

Ultimately, we want to use those APIs to write memories to a prospect's mental hard drive, if I can put it that way, so when they enter a buying situation, their initial mental search already includes our product. (In AI terms, we can say it's a bit like context engineering for humans.)

The key is, of course, to know that there are in fact two different attention APIs at work, and that each requires its own distinct communication technique, hence the importance of understanding our two story types: story 1 *and* story 2.

Your turn

Here's a quick diagnostic test. Open a note or pick up a pen and paper and give it a go. In your case:

- Why should a buyer pick the *category* of solutions you're in for their problem?
- Why should a buyer pick your solution over the competition?
- What simple message should they remember and associate with you and your product?

Seriously, write it down. Done? Ok, now look at your homepage. Does that match? What about your sales deck? What about your LinkedIn presence or marketing campaigns?

What's the consistent takeaway from all of those messaging surfaces? That is, what's the final brand memory you're hoping to form in the prospect's mind?

Memetic hook

That brand memory is the final destination for our positioning work. That's the memetic hook we're ultimately trying to deliver.

In tech, we often talk about products going "viral," but we don't talk as much about how ideas themselves fundamentally spread, even though we all love "memes."

In fact, I'm often surprised how, when people talk about narrative and positioning, it's not the final deck or messaging that matters so much as it is something far simpler — a memetic hook that leaders can use with their team and the market.

On the leadership side, for example, Stripe's mission statement proved to be a rare case of a mission statement with actual meaning. They wanted to "increase the GDP of the internet," and that proved to be an oft-repeated memetic hook that their employees rallied around.

On the market side, that's the goal of much of what passes for category creation — to seed a memetic idea in the market that points back to your product. In fact, "category creation" *itself* is a memetic idea that was deliberately seeded in the market by a consultancy trying to point VCs and founders back to them. And it worked!

We evolved for many thousands of years with oral culture as the sole means of passing down information, and our brains are well-primed to send and receive these sorts of catchy packets of cultural information.

In a way, that's what all this work boils down to. That's why I coined the term "super positioning." That's why you'll inevitably work so hard to simplify and concentrate your own ideas: they need to spread.

Well before internet memes took over, Richard Dawkins coined the term "meme" to describe this kind of *cultural* evolution (as a riff on gene), which philosopher Daniel Dennett later described as "information packets with attitude."

I love that as a description of the ideas we're often trying to seed in the market and the position we're trying to build in the buyer's mind — little 'information packets with attitude' that form those all-important name/need memory associations. Need a website, think Squarespace. Need a CRM, think...

Exactly.

After all, brands that get remembered are brands that get bought. Buyers can't buy you if they've never heard of you, don't remember you, and don't associate you with their core problems. But the flip side is that the more folks you *can* hook with your message, the more you can sell.

And the more you can sell, the bigger you'll be.

What about category creation?

If super positioning is based on the hierarchy of attention and generating those memetic hooks, how does it compare to category creation? And why do I think it's dead?

Well, what *is* category creation?

Remember how we discussed traditional positioning being a mix of thinking as big as possible and as small as possible at the same time, and how confusing that can be, because they're a function of our two modes of attention?

And remember how we just saw that the right hemisphere tends to global and the left hemisphere focuses on what's local?

That same confusion plagues category creation, as we'll unpack next chapter. At the heart of the concept is a contradiction — the aspiration is to create a company so singular and unique it transcends categories, but the playbook and very name of the concept all point people back to hair-splitting sub-category names that usually just confuse buyers.

That fundamental confusion makes the concept redundant — we now have better tools and better language to understand what's going on.

But even if you buy into the premise of category creation, the more practical answer to 'Why is category creation dead?' is that category creation has been done.

There are already 2000+ B2B tech categories (thanks G2.com); creating the 2001st 'category' just isn't that interesting. It doesn't mean that it doesn't still happen; it does! It just means that *on its own*, a made-up category doesn't make for a compelling, catchy, memetic idea anymore, if it ever did.

Your category is still part of the positioning mix, to be sure, and I'll show you a much better and far more straightforward formula for figuring out your *concept* in our **strategy section**, but "category," for us, is just not a meaningful lens. It's redundant. It's dead.

Transcending categories

Nevertheless, founders, investors, and marketers *did* buy into category creation for a very real reason.

However flawed the concept may be, it gave people language to describe **an approach to startup positioning that was distinct from regular competitive positioning.**

When regular competitive positioning gets too niche focused, there's a risk that venture-backed founders and their teams get stuck building a low-stakes business. That might be a perfectly fine business, people may love the product, and you may be having a wonderful time working with a niche set of customers, but for better or worse, investors did not give you $1M, $10M, or $100M+ to build a low-stakes business.

They need you to be focusing on the biggest possible outcome instead, and for a while, that meant being told you had to "create a category."

That's the value of unique language — it gives people tools to describe and collaborate on unique goals that aren't captured by existing language. And the unique *need* here is that venture-backed startup positioning is not the same as regular competitive positioning.

That's why I think you should focus on creating a super position instead.

In super positioning, the biggest outcome isn't *creating* a category, it's creating a super position that *transcends* your category entirely.

There's search and then there's *Google.*

There's CRM and then there's *Salesforce.*

There's marketing automation and then there's *HubSpot.*

There's design tooling and then there's *Figma.*

There's productivity software and then there's *Notion.*

There's payment providers and then there's *Stripe.*

And so it goes: super positions transcend categories.

Super positioning itself doesn't aspire to be another kind of positioning either; it's an approach that transcends existing approaches by getting above them and mapping them to their respective mode of attention.

How do you create your super position? By finding ways to be super relevant, super useful, and super memorable, hitting both right and left-brain attention in such a way to create memories that stick.

Those companies above are now all epic brands in their own right. But before that, they were startups that had to start from modest positions to get the flywheel going. It's that end of the startup journey that we're focused on here.

The three prongs

What should you actually do to start working on your super position? I've got a very tight system for you coming up (Again, feel free to skip to the **hands-on chapters** if you want to get stuck in right away!) that matches those common positioning tools to the mode of attention you're working with. The aim is to provide a way to incorporate both:

- **Maximal outcomes:** A new way to think of maximal startup outcomes — companies that create a 'super position' in their market (not a 'category').
- **Minimal starting point:** A more grounded way of incorporating the path they took to get there (often starting in a niche, not naming a category).

That's the think big/think small nature of super positioning — it's not either/or, it's both!

To do that, we're going to draw from what works. Very smart people have done some fantastic work trying to nail down how positioning works, and there are plenty of insights I'll draw on in this book as we go.

I'm standing on the shoulders of giants who are vastly more influential than your humble author.

However, the confusing thing is, as mentioned, that different schools of thought seem to advocate for diametrically opposed positioning approaches — and often throw shade at the other approaches, too.

Theirs is the right approach (thesis), unlike the other approach (antithesis). But now, with our more sophisticated understanding of left and right-brain attention, we can construct a bigger umbrella (synthesis) for all these approaches. That's the joy of super positioning. (And dialectics, I guess).

In the next chapter, we'll look at the first prong of super positioning — narrative and big change stories.

5

Prong 1: Go big(ger) — categories & narratives

The first prong in super positioning is change stories, and the biggest proponent of the biggest kinds of change stories — category creation — was 2016's *Play Bigger*, as we touched on last chapter.

Play Bigger didn't invent category creation, however.

It was first popularized by the two positioning OGs — ad-men-turned-strategy-consultants Al Ries and Jack Trout.

(Ries & Trout popularized "positioning" itself in the late '60s and wrote the seminal book on the topic: 1981's *Positioning: The Battle For Your Mind.*)

In 1985, Ries & Trout alluded to category creation in their book *Marketing Warfare,* writing "A flanking attack often hinges on your ability to create and maintain a separate category."

By 1993, they'd turned it into one of their "laws", writing in *The*

22 Immutable Laws of Marketing that the second law was "The Law of the Category: If you can't be first in a category, set up a new category you can be first in."

Eddie Yoon would briefly pick up the idea again in 2011 (and 2013), writing a short piece for HBR on why "Category Creation Is the Ultimate Growth Strategy," paving the way for what was to come by labelling innovative businesses as "category creators."

You can see the 2011 blip on this Google Trends chart:

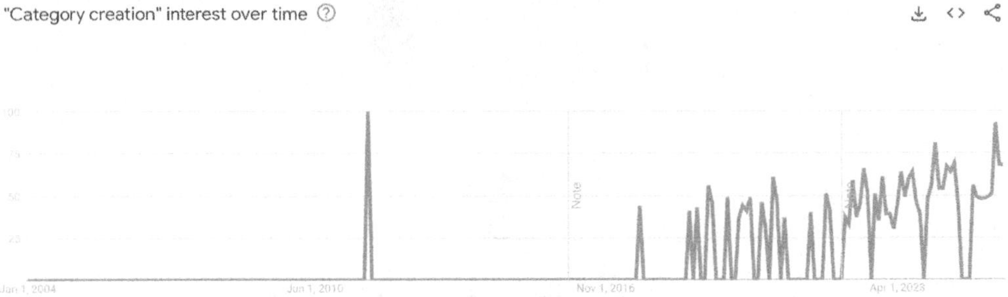

But otherwise, it was all quiet on the category creation front.

Then *Play Bigger* happened.

Play Bigger turned category creation up to eleven. The authors — Al Ramadan, Dave Peterson, Christopher Lochhead, and Kevin Maney — weren't shy of a big change story. They claimed their approach superseded all that had come before — including Ries & Trout.

In fact, the *Play Bigger* authors humbly submitted that, in the great tradition of American pop business books, their theory of category creation explained not only all business success ever, but indeed more or less all human success ever, from Google to Gandhi. (By the end of the book, you're left wondering if the Big Bang was the original category creator because it 'created the category' of a habitable universe.)

Unfortunately, it turns out there isn't much you can learn about B2B SaaS by trying to draw a line between Elvis, Paul the Apostle, Lenin, Peter Drucker, and Salesforce.com. (Yes, that's really what *Play Bigger* does!) But that didn't stop them trying.

Here they are introducing their own category (so to speak) in *Play Bigger*, the concept of 'category design':

> *Category design isn't just positioning or branding. We have great respect for the seminal 1970s book Positioning, by Al Ries and Jack Trout. Their sense of positioning as a necessary discipline was on the mark for the late twentieth century. But in this century of mobile/social/cloud, dynamics have changed, and positioning is only one part of category design. And as for branding—we call branding agencies "tattoo parlors." You don't want to wake up with Mike Tyson's tattoo on your face, so don't let someone else tattoo you. Brands don't make a category king. Category design does.*

I think the book makes as much sense as the Mike Tyson tattoo riff above (what?). I think we deserve serious frameworks grounded in reality, not a fever dream that sounds like the start of a corny joke. (A Christian apostle, a Marxist revolutionary, and a SaaS CEO walk into a bar...)

Nevertheless, I do have some sympathy for the need they were trying to address

That's why, with super positioning, rather than taking a top-down approach of trying to find some commonality between Everyone Who Ever Did Anything Great Ever, we take a much simpler approach of working from the bottom up from our brains and how they attend to the world.

Here, however, I want to get a little meta and analyze *how* the *Play Bigger* authors deliver this message, because it's one heckuva change story and the result was very impressive — they successfully meme'd an entire industry (venture capital)!

The change story

If you go back over the quote above, you'll quickly recognize it as a big story 1 pitch. Positioning (of the Ries & Trout variety) is dead! Branding is bad! You need to 'play bigger' because *so much has changed!*

That is, they're making their case for why their approach is now super relevant due to all this change. Story 1 pitches are very much about external change over time, and the *Play Bigger* folks roll everything internet (mobile/social/cloud) into a single bundle of change. That change, they argue, demands a new approach, and they've got just what the doctor ordered: category creation and 'category design.'

While I think "category creation" is fundamentally a bad metaphor, again, I get why people liked the concept. It seemed to speak to everything that's exciting about startups and tech — big innovation, unique business models, and once-in-a-generation success. That's exciting to anyone interested in business. Even if your reality is much more mundane, it's fun to dream about those kinds of outcomes!

The problem, however, that I and many others have noticed, is that everyone who reaches for category creation seems to grab the wrong end of the stick. I've seen people get tied into knots and waste months and years on fake "categories" that go nowhere.

It's conceptual quicksand that sucks folks in — especially well-meaning CMOs and marketing teams — because it *seems* like the highest-order bit to be working on, especially if you're a startup, even though it's often a dead end.

But it illustrates a powerful point about how we understand the world and the memetic hook you ultimately want to build as part of your super position.

Metaphors and misunderstandings

In *The Master and His Emissary*, McGilchrist writes:

> *Metaphoric thinking is fundamental to our understanding of the world, because it is the* only *way in which understanding can reach outside the system of signs to life itself. It is what links language to life.*

That is, metaphoric thinking is how we understand the world. But metaphor is also how we *misunderstand* the world, and category creation, as a metaphor, throws people off in a way that cuts right to the heart of our two modes of attention.

Ironically, it can be read in a very metaphoric, right-brain way and a very literal, left-brain way.

When one term can be read in two very different ways, it will inevitably leave a lot of folks confused, and that confusion can send your company on a wild goose chase that's a distraction at best or a huge missed opportunity at worst.

First, though, here's the more metaphoric right-brain understanding.

When strategy consultants and investors talk about category creation, their emphasis is very much on the *creation* side. The implicit message is you should create something so unique, so innovative, and so fundamentally *different* (not merely *better*) to everything that's out there that *the outcome* is that your product or company — in a metaphoric sense at least — needs its own category.

There's a lot of right-brain attention at work there. Not just in the fact it's metaphoric, but what's being evoked — change stories, what's new, what's 'out there' on the horizon.

I like that!

The problem is that most folks don't get that implicit message. Instead, they hear the explicit "category" part of "create a category" and take it very literally. As in, you need to pull a new literal category name out of thin air for the software or product you're making, spend months and years marketing that, hope it catches on, and then hope that, somehow, people buy your product. That's what *Play Bigger* explicitly recommends, and that's where teams often drive straight into a positioning ditch.

This is the more left-brain approach, where the focus is on the map and, well, the literal categories. It's a product of laser-beam focus on how we split the difference between things.

Even the best get confused by this. Brian Halligan, co-founder, former CEO, and now executive chairman of HubSpot, wrote in 2024 that (his emphasis):

> ***Creating a* category *is harder than it looks*:** *HubSpot*

> *created the "inbound marketing" category. Pulling that off involved writing about a zillion blog articles, giving a jillion speeches, writing a book, running a conference, etc. We invested way more energy in creating the inbound marketing category in the early years than we did in marketing the HubSpot product. ...So, when we wanted to go into the sales category, we thought we could just re-run the same playbook for "inbound sales." Failed. When we went into CRM, we thought we'd create a new category called CMR, "customer managed relationship" software. Failed. When we released our CMS, we thought we'd create a new category called COS, "content optimization system." Failed. In retrospect, we caught lightning in a bottle with "inbound marketing."*

I realize we're deep in the weeds here, but there's a profound point here that goes to the heart of why some "category creation" attempts succeed and others fail, and it can be hard to see it, even if you're one of the best to ever do it, as Halligan and HubSpot were.

The difference goes to the right-brain and left-brain ways of seeing things, as we've discussed in terms of innovation and labels. Consider the difference between HubSpot's attempts at:

- Inbound marketing.
- Inbound sales.
- Customer managed relationship (CMR).
- Content optimization system (COS).

Halligan calls them all categories, but it's the last two that are literal categories of software — riffs on existing corporate software (CRM and CMS).

Inbound sales and inbound marketing, however, aren't categories of software. ("Marketing automation" was the literal category of software for HubSpot's inbound marketing.) They're playbooks. They're what you do as a company or as a marketing or sales department. And what's the difference between inbound marketing and inbound sales?

Inbound marketing caught a wave — the organic digital marketing wave. That wave — originally the rise of blogs, social, and search — created the *demand* for a new playbook. Things had changed, so people were naturally receptive to a new way of doing marketing.

(That's why **riding a wave** is the first of our four super positioning strategies, and HubSpot is one of the main examples.)

In fact, many people were already doing it in different ways, but HubSpot were the ones who could put their finger on it, give it a name, create a playbook and solution for it, and market the hell out of it. (Halligan isn't kidding when he says it involved "writing about a zillion blog articles, giving a jillion speeches, writing a book, running a conference.")

Those other three category creation attempts? No wave; no "category" created.

The power is in the wave, as we'll discuss in **chapter 8**. It requires a great deal of astute right-brain attention to what's going on 'out there' in the world, and the result you need to aim for is a new *concept*, not a new category of software for its own sake.

This gives us our *concept = change + playbook + tool* formula that we'll explore in the strategy section, but for now, let's discuss *concept* creation.

Concept creation

One of the weird parts about category creation, as an idea, is that the kinds of businesses that are celebrated category creators (like HubSpot) don't necessarily create a new *literal category of software*. Google didn't create search, Facebook didn't create social networking, and Salesforce didn't create CRMs.

This is where the category creation metaphor misses the mark. A better metaphor, and what people really mean, is *concept* creation.

New business concepts — like Google ranking pages using links, showing 10 blue links, and not plastering their results pages with banner ads, or Salesforce selling their software over the web (weird!) at the turn of the millennium — can compound in such a powerful way that you become the dominant player in your category.

While pulling off an infinite cash generation machine like Google is truly one of those once-in-a-generation feats, the idea of pursuing a singular business that makes unique choices is business strategy 101.

Here's Joan Magretta from her excellent book *Understanding Michael Porter*, for example:

> *Managers often think about competition as a form of warfare, a zero-sum battle for dominance in which only the alphas prevail. This [...] is a deeply flawed and destructive way of thinking. The key to competitive success—for businesses and nonprofits alike—lies in an organization's ability to create unique value. Porter's prescription: aim to be unique, not best. Creating value, not beating rivals, is at the heart of competition.*

Nothing wrong with that advice. Startups are meant to be ambitious after all.

Ok, so businesses should compete to be unique, as Porter says. I like to say they should create a new *concept* not *category*, necessarily. Isn't that a kind of pedantic point?

Pedantic? Me? Surely you jest! But we do need to do a little more untangling before we get to the bottom of category creation, because it shows why super positioning is so important.

Enterprise concepts

Category creation discourse often muddies the water between the mega winners of the internet revolution (mostly B2C companies like Google, Facebook, Netflix, LinkedIn, etc.) and the decidedly less sexy world of selling concept-driven enterprise B2B software (we're pioneering the GEO category — *SEO for AI!*).

The funny thing is, there's a better metaphor for those B2C companies (Google, Facebook, Netflix, etc.): *aggregators*.

(Yes, I'm still picking this category creation bone, stay with me!)

Ben Thompson, who writes the excellent Stratechery, coined Aggregation Theory back in 2015 to describe these companies who succeeded thanks to their ability to *aggregate* the infinite content the internet enables.

By aggregating it in one place (Google for the web, Facebook for social connections and then social slop, Netflix for media originally, LinkedIn for professional connections, etc.), they become increasingly useful to both creators *and* users, which makes them incredibly sticky (and profitable!) businesses.

In B2B — and this is my thought, not Thompson's — you can think of category leaders as *aggregators of industry demand* that they solve with their platform, ecosystem, and support services.

That's why incumbents are so hard to dislodge and why customers naturally gravitate towards them — the incumbent (and its vast ecosystem) is more likely to have come across, and created a solution for, whatever unique needs the customer has, even if it's at the cost of product purity. (That's why "enterprise software" isn't exactly synonymous with "delightful experiences.")

The more they've aggregated, the lower the risk for a customer to buy them, and the easier it is to get a deal done.

Aggregation Theory is a fantastic framework that comes from the Michael Porter lineage of positioning as business strategy, where we can analyze where a business sits — i.e., its position — in a value chain that runs from supplier to end user or customer. It has *far* more explanatory power than anything to do with "categories" to explain the success of those firms, and yet the much more crude "category creation" caught on in the VC world.

But *why* did "category creation" as a term win?

Because it had a super position!

The 'problem' (so to speak) with Aggregation Theory is that Thompson had no need to market the theory beyond promoting it to his audience for their intellectual edification.

Category creation, on the other hand, is a product of consultants (those pesky consultants!), and the *Play Bigger* folks marketed the hell out of their theory and created the strongest position.

They had the big narrative, the specific competitive playbook, and brand message repetition that created a name/need association between "category creation" and "big winning startup," as I touched on earlier. That's what creates a super position. (That, and the endless confidence to hammer everyone over the head with the concept.)

The best product — or theory, in this case — doesn't always win, and it was ever thus.

Meaning wins

These ideas only stick if the latent demand is there. For an idea like category creation to take on memetic proportions and basically meme a whole industry (venture capital in the late 2010s), there must be some strong demand for it to tap into, and beyond language, which I touched on earlier, one of the biggest sources of demand is the need for *meaning*.

Play Bigger succeeded for a time because the language of category creation gave founders and marketers a greater sense of meaning in what they were doing.

They weren't just doing boring ol' B2B marketing anymore (**yawn**), they were *creating a category!* They weren't just marketers, they were *category creators!* That's not just category creation, that's *identity* creation. There's a whole pre-packaged identity there waiting for you if you buy into this world — one that differentiates you from your behind-the-times peers.

That's a lot of meaning. You and I might find it all a bit eyeroll-inducing, but there's a lesson in there, especially for B2B. People — newsflash — generally don't find their white-collar work to be super inspiring or meaningful. If you can give them an identity and narrative to adopt that *does* make their work seem more meaningful, then that increases the chances they'll adopt your concept. And that's what worked for category creation and *Play Bigger.*

Maximal ambition

In super positioning, we aim to dial this kind of story 1, right-brain new change/new way attention up or down depending on the context we're operating in. But what did dialing it up — *way up* — look like for the *Play Bigger* folks themselves?

Well, if we study what these consultants *did,* not just the explanations of *why* they did it, an interesting pattern emerges. I think of this as the meta *Play Bigger* playbook, and it looks like this:

- Coin a new phrase (such as 'category design' for their own work).
- Declare it a new discipline.
- Describe a playbook for that discipline.
- Suggest the company go all-in on it.

How did this play out? Here are a few examples from the book on Macromedia, Qualtrics, and *Play Bigger* itself), along with a more recent attempt with B2B SaaS ~~whiteboard~~ "visual work platform" Mural.

Across these case studies, we see:

1. **A simple concept name:** 'Experience design' (Macromedia), 'Category creation' (Play Bigger), 'Experience management' (Qualtrics), 'Collaborative intelligence' (Mural).
2. **'A new discipline':** For Macromedia, "experience design" would be a "new discipline" that would be "part of the fabric of the technology industry"; for Qualtrics, experience management would be a "new business discipline"; for category creation itself "we believe it's time to develop a new *discipline* called category design" (their italics); and for Mural, they would introduce a new discipline called "Collaboration Design".
3. **A new playbook:** You can't have a new discipline without a new playbook, and Macromedia suggested Rich Internet Applications (or at least a web design suite of tools) was the future; Qualtrics had a specific view on closing the "experience gap" ("the difference between what a business believes is happening and what's actually happening"); *Play Bigger* was, in part, a literal playbook for market strategy; and Mural had a playbook they acquired (LUMA, and their

"LUMA Institute") that was supposed to result in the "Collaboration Design Institute™."
4. **Going all in:** Macromedia didn't just float a concept, they put their money where their mouth was with an (at the time) bargain bundle of tools; Qualtrics made "XM" (for "experience management") a core part of their brand (even their stock ticker was $XM); *Play Bigger* claimed anything and everything was category related, stretching their concept further than perhaps any business concept has been stretched before; and Mural ended up with, well, concept soup: Collaborative Intelligence, Collaboration Design, LUMA, Institutes/CDI, and that's before we get back to the actual product: Mural.

(Quotes without links are from *Play Bigger*, and while I'm cherry-picking examples a little bit here — sometimes they did just focus on literal categories of software — I think you get the gist.)

Note the concept focus here. This new terminology, (supposed) new discipline, new playbook, and new branding all exist *around* the product.

It's an approach that, at its best, is about innovating at the concept level, not the technology or product level, per se. Ironically, it doesn't have much to do with literal categories of software, either.

It's also highly B2B-centric. The great flaw at the heart of *Play Bigger's* often-absurd category creation claims is that the playbook outlined above has absolutely nothing to do with their choice of "category creators" (Google, Facebook, Netflix, Elvis, Gandhi, etc.). They didn't follow *any* of these steps around new concept names, disciplines, and playbooks. The actual playbook here is highly B2B-centric, which is why B2B investors ran with it.

In super positioning terms, however, we can understand what's going on. We can see that, in practice, this is about innovating around what you put on buyers' right-brain radars. It's *not* about the product-level innovation they get hands-on with. It's about the big ideas for execs and leaders that gives them a new way to think about what they're doing.

Or so the theory goes.

Occasionally, in the B2B SaaS world, this works a treat. Qualtrics, with their XM or "Experience Management" concept, is probably the poster child there. Huge respect for that work, which we'll unpack a bit more in subsequent chapters.

Sometimes, however, it flops. Mural's "Collaborative Intelligence" category has been scrubbed from existence for example and, to be honest, just wasn't that meaningful. But hey, can't win 'em all.

Description not prescription

I should be clear that I'm not prescribing the *Play Bigger* approach here — we'll look at a much more concise, concept-driven approach when we dig into wave riding in **chapter 11** — but I want to state that, as much as the word nerd in me might dislike category creation as a metaphor, there are useful tools there, and if they work for you, great.

Alright, that's that bone thoroughly picked. I hope it at least helped illustrate what the maximal right-brain, put-something-new-on-their-radar approach looks like.

It's not the only approach at this end of the spectrum, though.

Another take on this kind of change story comes from Andy Raskin, famous for his strategic narrative methodology, which follows a similar story 1 pattern.

The strategic narrative

In **chapter 9** we're going to get hands-on with you writing your own strategic narrative using the super positioning approach, but we'll just touch on the classic approach here. To paraphrase Raskin:

- There's some big change 'out there' that's external to your company.
- It's high-stakes, do-or-die stuff.
- There's a promised land on the other side.
- Our tools and features will help you get there.

When a narrative literally starts with change that's happening 'out there', you know you're starting with a very right-brain approach to your story.

Also, isn't that what we just saw with the category creation folks? I don't mean in terms of ideas, I mean in terms of rationale — there was a big change (mobile/social/cloud), it's high stakes with winners and losers, there's promised land ('category kings'), and tools to help you get there ('category design').

I mean, kind of, yes! I'm not suggesting any relationship between the two at all, but it shouldn't be a surprise that similar approaches to change stories exist, **because they're all trying to target the same underlying mode of attention** — the right-brain radar.

That's one of the universal parts of super positioning — these right-brain, story 1 style pitches. Raskin, to his credit, articulated it very concisely in his strategic narrative approach.

You can see how, in both approaches, there's a *lot* of story 1 context-setting before the shift down to any story 2 specifics around a product and its features.

Category creation focused on the name of the *category*.

Raskin's strategic narrative focused on the name of the *change*.

In super positioning, either can work; the point is just that you have a *concept* that ultimately helps you transcend your category.

That might seem like a nuanced point, but it's not. The metaphor you choose matters. The metaphor *you* choose for your own product and playbook matters, too.

Competitive positioning

This is big right-brain stuff, after all — change over time, the new, the global, the opening up to new possibilities, and the integration of a new reality.

But what happens in the left hemisphere when it comes to competitive positioning? That's what we'll explore next.

(Want to get hands-on with your strategic narrative instead? Jump to **chapter 9** to explore narrative or **chapter 11** to look at right-brain strategies and examples.)

6

Prong 2: Go narrower — niche, analysis, & vision

On the one side, we have these story 1 pitches that are, on the whole, usually about a concept that's bigger than your company or product on its own. They might be about a change happening 'out there' on the radar that necessitates a new way of doing things, for example. Done well, that drives relevance.

On the other side, we have the story 2 approach. This is the world of Geoffrey Moore (*Crossing the Chasm*), April Dunford (*Obviously Awesome*), and the Jobs to be Done theory (especially Tony Ulwick with *Jobs to be Done: Theory to Practice*).

This is about zeroing in on usefulness — *and to whom* your product is uniquely useful.

Moore advocated building momentum in a niche with tight positioning so you could 'cross the chasm' between early adopters and the mainstream.

Dunford gave us a positioning methodology for finding best-fit customers who care a lot about your unique value. (I am, for the record, a huge fan of April and wouldn't be writing this if it wasn't for *Obviously Awesome* and her tireless work advocating for positioning.)

And the Jobs to be Done theorists and practitioners like Ulwick have been describing and debating measures of usefulness, i.e., how well a product serves a given customer's 'jobs' that they're trying to 'get done' for years.

We'll unpack this prong further when we look at our 'find it' strategy in **chapter 12**, but here I just want to give you a taste of the second prong of super positioning — the world of left-brained competitive positioning.

If the right-brain story 1 approach is looking up and out at a global change that makes your product relevant, the left-brain story 2 approach is looking down at where you're most useful.

Looking down

Dunford's approach, for example, is interesting precisely because it's not so much about finding your positioning 'out there,' it's mostly about finding it *down* in your existing customer base.

Here's Dunford in her positioning classic, *Obviously Awesome*:

> *[A fellow repeat VP of Marketing and I] discovered that our philosophies were strikingly similar—we spent all our time looking for things that work, and then deciphering why. What marketing campaigns brought in the most leads? Which pieces of content were consumed the most? What events had the most attendees? Asking these questions led us to the happiest customers. Once we narrowed in on the happiest customers, we went looking for the reasons they were so satisfied while others were not. What was it about our offering that made them so happy with it, and what was it about those customers that made them such a good fit for us? Answering these questions helped us figure out what our value was, who it was resonating for and why—in other words, it helped us get a*

start on understanding what our positioning should be.

The key in this quote is that Dunford "narrowed in" on the happiest customers. If you're narrowing in, you're using left-brain attention.

That is, if a lot of right-brain story 1 pitches are about what *could be*, this more left-brain story 2 approach is about 'narrowing down' to see and analyze what *is*.

This is the left-brain laser beam in action. It's not just about "talking to your customers" in some generic get-the-gist-of-what-people-want sense, it's about doing the proper analysis into what works, for whom, and why.

Finding your market vs. mirroring your market

This left-brain approach to positioning is often much more structured and methodical than the old "talk to customers" cliche. The advice to get out of the building and talk to your customers is obviously good advice — it's hard to build for and sell to a market you don't fully understand — but what are you actually looking for? What do you hope to see?

Ultimately, what you want to see is that you've become a 'virtuous mirror' of your chosen market, if I can put it that way. You're the solution that perfectly reflects their problems, your marketing reflects the exact language they use, and you help your customers in a way they feel good about, so they tell others about you (that's the virtuous part).

That requires a broad and deep understanding of your chosen market.

But how do you choose that market? This is where the second prong of super positioning comes into its own.

Average vs. outlier customers

Successful competitive positioning often requires doing the work to find that segment or market to mirror in the first place. And that can require finding the outliers among your average customers.

Perhaps you've already seen the problem you're solving first-hand so there's not much question about which market you serve

— if so, that's fine, feel free to skip ahead.

But other times, lucrative positioning opportunities emerge because someone found the *outliers* who were getting outsized value relative to your *average* customers.

Indeed, one of Dunford's famous positioning stories is about a database product that flops, she rings around to see if anyone would care if it went away, finds a single-digit percentage of outlier customers who *are* getting value out of the product, they completely reposition around that, and the product goes on to have a very successful life.

If you "talk to customers" and take the average, you would say the product is indeed a flop and should be killed off.

But if you "talk to customers" and find the tiny sliver of successful outliers, then you can open some interesting doors!

We'll discuss this more in our **'find it'** strategy chapter, but whenever anyone suggests that you can fix your positioning by talking to your customers, you need to diagnose whether your problem is (a) your current market choice is correct, you're just struggling to communicate to that market, or (b) your current market choice isn't working, and you need a new vision for the product, which needs to come from either the founder's vision (i.e., a new story 1 narrative) or your collective *customer* vision (more story 2 found value), which we'll discuss below.

Main character energy

From time to time, Dunford will call out marketers on LinkedIn for getting carried away with what is essentially story 1 narrative, criticizing "marketers [who] can get a bit of main character energy when it comes to this stuff." In Dunford's view, "our goals as vendors don't belong in the center of our positioning. The value we can deliver for our customers needs to be the core."

That is, narrow your focus to the specific value that you can deliver, and stop worrying so much about the change-over-time story (especially, she says, if you think you're starting a 'revolution' or a 'movement').

This gets to the heart of the conflict between the right-brain, story 1 school of positioning vs. left-brain story 2 approach.

On the one hand, marketers absolutely can get carried away with their change story. Plus, as we'll see later in the book, there are plenty of folks who *found* their winning position without *~~creating a category~~*, for example.

To be fair to Dunford, her advice is not that you should *never* create a category; she just thinks it's rare — maybe something 1 in 10 companies should do, and her focus is much more on finding patterns in your existing customer base.

Average vs. outlier startups

This is where it's critical to know which game you're playing.

The question isn't go big or go niche, it's how do you generate the momentum necessary to reach the goals you've set?

Obviously, for early-stage venture-backed startups being the rare outlier — the 1 in 10 — is ultimately the point.

If every startup could just build a product, click their fingers, and make that happen, they would. But obviously it's not that easy!

So maybe for 9 in 10 startups, the answer is to narrow down and narrow down until you find a segment, however small, where you can be super useful and start building momentum up from there. Think of it as the minimum viable niche to match your minimum viable product.

But you do that because you have to, not because you necessarily want to. And occasionally, maybe the wave sweeping the market is driving relevance to such a degree (like the 2020s AI wave) that riding a wave is the more obvious way to build momentum.

Either way, the point is to find momentum and be the outlier. After all, venture-backed startup positioning is not necessarily like other positioning — you *have* to be working toward the huge home run that could return the fund.

Ideally, in super positioning, you do that by using the strengths of both modes of attention for the best possible outcome. That means having a strong *company* vision that works hand-in-hand with ongoing *customer* vision.

Company vision vs. customer vision

Our brains, and our attention preferences, dictate what we see, both in a metaphoric and very literal sense sense.

This is where the hierarchy of attention gives us more gold — it helps us understand how vision works, too.

We talk a lot about 'vision' in tech, but we rarely define what it is. I touched on the story 2 approach to positioning above being a function of your *customer* vision — what you see when you look down at what people are actually doing.

Consider how that contrasts with *company* vision in startup land which is focused on looking outward at what might be out there on the radar. At the big end of tech town, those visions have expanded to encompass the entire universe itself, which is a bit of a worry!

The light cone of all future value

Company vs. customer vision, like their respective sources of attention, can often be at odds. Those with the big vision don't want to look down at what is because it might puncture their ego; those looking down at what is don't want to take the risk to imagine what might be.

But ultimately, they need to work together and intersect in reality if you're going to have a chance of making it.

That can be a challenge when the industry gives story 1 'vision' near messianic reverence to the degree that, in other industries, folks might start questioning your sanity.

For example, if you're Sam Altman, in 2019, three years *before* (!) ChatGPT's breakout success, you could say that OpenAI might "capture the light cone of all future value in the universe," and people would then throw billions and billions of dollars at you on the back of your, shall we say, rather unique vision.

And, in Altman's case, even if you miss the moon, you still end up with one of the most valuable startups on the planet... in part because the product turns out to be super useful *and* memorable because you win the consumer brand lottery, of all things.

That's tech, baby.

We're focused on B2B here, and while those once-in-a-generation visionary startups might get all the headlines, the day-to-day for most folks is trying to figure out what's working and what they can position around.

Visionary buyers & investors misfires

Part of that positioning work can, in the early days, mean finding your "visionary buyer." Remember, customers use these same two modes of attention, and that can manifest in different ways.

Sometimes it's buyers that are the ones looking out on the horizon for that early tech that will give them an edge by being super useful to them, bugs and all, and sometimes the startup has to use their laser beam to find them.

B2B event marketing platform Goldcast found their visionary early adopter in Drift's then-CEO, David Cancel. Palash Soni, Goldcast's CEO, in a discussion with B2B SaaS founder and educator TK Kader, mentioned the "massive bet" Cancel took on their platform, and that's exactly what an early, visionary customer does. They take a risk on your tech, and if it goes well, you get their endorsement.

That's great, but what are customers — including 'visionary' ones — buying?

Soni and Goldcast also ran into the story 1/story 2 trap with their VC's right-brain obsession and their customers' left-brain focus. To quote Soni (lightly paraphrased):

> *VCs only care about visionary stuff, right? [...] So we only talked about 'the message' and 'category creation' [...] but we were so wrong about it [...] because the customer is buying features ultimately.*

So they focused much more on story 2-style product marketing instead of the story 1 narrative and were much better off for it.

This is very familiar to the Geoffrey Moore *Chasm* playbook — finding those visionary customers, selling to a specific social group (event marketers, for Goldcast), and establishing leadership in a niche with a strong, capable product. That approach gave them the foundation to catch the AI wave, and now they're leaning hard into B2B video content creation.

If you can get your story 1 *and* story 2 pitches right, you end up with happy investors (Goldcast has raised $42M) *and* happy customers. If you get it wrong, you're pitching features to underwhelmed investors and vision to confused prospects who don't buy.

Customer vision & enterprise opportunities

Customer vision isn't always about methodical, backwards-looking analysis. It's also about having an idea, putting it out there, and staying acutely aware of emergent opportunity.

There's a startup joke from Dickie Bush that goes:

1. *Start literally any business*
2. *Find 1 part of starting your business that sucks*
3. *Create a new business to solve that problem*
4. *Sell your new solution to business owners who have that problem (and are willing to pay for it)*

Except it's not a joke, he's being serious, and for good reason. For example, maybe you start working on a quirky MMORPG, build a bunch of stuff, and eventually see the opportunity in the chat functionality you've created, and turn that into a $27.7B exit to Salesforce. (That's Slack.)

Or maybe you're Whatfix, who have managed to raise $280M despite their very small and scrappy start.

As Avinash Raghava's documents in his piece on Whatfix for *The Grit Stories*, Whatfix's founders first built an SEO tool for SMBs, but churn was high due to the complexity of their recommendations.

They then added a 'fix it' feature, which guided users on 'fixing' the problems the original tool found. That 'fix it' button turned out to be a more promising opportunity — SMBs wanted to use it internally — but the SMB world was still a slog. Growth was slow; churn was high.

That little button would turn out to be super useful in the right context, though, and the art of this story 2 approach to positioning is making sure you are in the right category.

Whatfix then looked upmarket and pioneered the emerging "digital adoption platform" (DAP) category. A DAP is software layered on top of existing enterprise software that helps with user onboarding, support, and performance.

CRM is 'low-hanging fruit' for DAPs, and on the *100x Entrepreneur* podcast, Whatfix founders Khadim Batti and Vara Kumar Namburu described their journey, including a bet they took as a tiny company when they decided to attend Salesforce's Dreamforce conference in the US.

They go to Dreamforce to further validate their market, and at this point, they'd been selling their product for $2,000 to $3,000 per year (i.e., $250/mo at the high end). At Dreamforce, they were aiming to pitch the product at the "expensive" price point of $8,000 to $10,000 annually instead. Makes sense, right? They're pitching enterprises, after all.

A prospect comes along, loves the concept, and asks how much it costs. The rep says "$8,000," meaning per year. The prospect does some quick mental math and says, '8 by 12... $96,000, that looks fine.'

That was, as you can imagine, quite the shock to the team. What they thought was an "expensive" annual price turned out to be an entirely reasonable *monthly* price for this kind of buyer. But the prospect was spending $20 million already, so an extra $100k on top for Whatfix was fine if it helped them get ROI.

The Whatfix team didn't just 10x their prices (or 32x their original $250/mo pricing!) and waltz into enterprise deals, though. There's no free lunch — there was still a lot of learning ahead. But by getting out of the building and talking to their customers, they stumbled across the fact they could close deals for $96k vs. $3k, and learned how to price and position their products. That opened up a whole new market opportunity, and as of writing, they have 10 million users and 700 customers, including some of the biggest enterprises on the planet.

Think of the kind of analysis, vision, and risks they had to get there, though. First it was identifying the right feature, then the right category, then the right buyer. But it goes to show you what can happen if you can bring all those ingredients together — you find where you're both super relevant and super useful, and enterprise buyers will pay accordingly.

The L in R/L/R

I really love this left-brain approach of analysis and putting the pieces together. It's where I started. But eventually I had to accept that attention should ultimately work together — we receive new information on the right hemisphere, process it on the left, and reintegrate it on the right, after all. (That's the simple model we're working with, anyway.)

This means that whatever you find with your left-brain laser beam, you're always looking for something you can put back on your customer's right-brain radar, if that makes sense.

Maybe that *is* a big, super relevant change story, and it works, and you mirror the success of HubSpot or Drift, as we'll see in our **'ride it'** chapter.

But maybe — and perhaps often! — your starting position in the market isn't the one that makes you successful. If it isn't, you need to deploy this second prong of super positioning and use the left-brain laser beam to find what's super useful, and for whom.

Maybe you find a niche to get started in, like Goldcast, and you focus on what makes you useful to customers, not relevant to investors. Or maybe you find a feature that works, and rethink your category, aspirations, and pricing, like Whatfix.

Then you take that, and put *that* back on your buyer's radar. It always comes back to the specifics that are super relevant, super useful, or both for the buyer you're targeting.

I call those specifics your "strategic specifics." Maybe you start with them and they work; maybe you have to find them. Either way, you should absolutely write down what yours are before putting them into the super positioning form of a strategic narrative, which is exactly what I'll walk you through in two chapters' time.

Maximizing customer vision

Super positioning depends on you using your left-brain laser beam to 'sample reality' — to see what customers are really doing and how they're really responding, and then bring that information back to bring your internal narrative so you're in alignment with the real world.

That's the art of competitive positioning, and that's why it's the second prong of super positioning. It's what brings you — and keeps you — in sync with reality.

If you can combine that with our first prong — a big change story — you can build an incredibly strong position in the market.

Rebooting the company around AI

Both modes of attention need to work together. A true visionary doesn't just see what's out there on the horizon, they combine it with an intimate knowledge of their industry and their customers' needs.

What happens when you really crank up both kinds of vision?

You can save your company.

The AI wave has given every company a chance to renew their vision, and that includes the big SaaS darlings, too.

Intercom's founders, for example, originally saw an opportunity to improve the support experience with a little messaging widget for your website (like an intercom, get it?).

That feature has since been copied to death — the meaningful uniqueness was gone — but it was an super useful innovation that took them to great heights in the 2010s, but their relevance was fading.

Long story short, they never went public, their co-founder CEO left, their growth stalled, and competitors caught up. They were suffering a severe lack of vision and resorted to squeezing every penny out of their customers.

That same CEO, Eoghan McCabe, eventually came back, reset their company vision on an aggressive AI bet, cracked the whip, repositioned the company around their original support niche, fixed their pricing, and went all in on AI in a massive way.

They weren't just "powered by AI" like everyone else, they

created a new AI agent business inside the company — Fin. A new product, a new name, with new tech, built on new economics, and a new vision *and* focus.

All of which dramatically restarted Intercom's growth, despite the fact they were already doing nine figures in revenue.

What a difference a strong company *and* customer vision makes! That's what competing to be unique is all about.

Super memorable

The point of carving out a unique competitive position isn't just so you can sell in the present. It's so you can build memories in the market — and the mind of the buyer — that will drive your GTM strategy beyond your marketing and outreach alone, too. Next, we'll look at brand memories, the third and final prong of super positioning.

(If you want to skip ahead and think through what you've seen with your customer vision, jump to **chapter 8** where we'll explore your "strategic specifics" or if you want more examples, check out **chapter 12** where we'll look at more examples of finding unique value in a niche.)

7

Prong 3: Go broader — brand, synthesis, & 2x2s

Maybe Dunford's approach resonates with you if you lean towards the left-brain mode of attention, or maybe you actually prefer the big right-brain category story. If you do, that's fine — venture-backed startups are about becoming the megahit, after all — but getting there *can* require going narrow first.

Those are the two broad B2B approaches that reflect our two modes of attention. But remember, there are three prongs to super positioning, and the third is memory. This requires giving our framework a bit of a B2C twist.

I don't think about you at all

Here's the gist. Some highly influential marketing academics think (with an amusing degree of left-brain certainty) that whether you go for the big change story or the narrow one, either way, you're flat-out wrong.

Customers, in this view, are "uncaring cognitive misers" who don't care about your story 1 *or* story 2.

In fact, to channel Don Draper, they don't think about you at all.

Any marketer, in this modern brand world, who thinks they can earn the loyalty of a set of customers by segmenting, targeting, and positioning around them (the "STP" approach) is fooling themselves.

The evidence, they argue, says the opposite — it's the bigger brands that reach more people (especially light buyers in B2C) that earn more loyalty. Not that customers are especially brand loyal anyway — if you want to maintain your brand memory, i.e., your position in the buyer's mind, you'd better keep advertising consistently and broadly, or you'll soon be forgotten.

Therefore, in this world of B2C mega brands, positioning is out. So is differentiation. And persuasion, message comprehension, and appealing to reason — in the bin, all of it!

This is the brutal world of Professor Byron Sharp's 2010 book *How Brands Grow,* which is one of the most influential modern books in mainstream B2C marketing. Applying those principles *alone* to the world of B2B tech sales would obviously be *highly career disadvantageous*, to put it mildly.

B2B != B2C

The key difference in B2B is that ACV can be very high, sales, as a function, can therefore exist, and you actually *can* then narrowly segment, target, and position for a specific set of customers, especially if you're a startup trying to break into the market!

The B2C marketing folks like to talk about marketing as a science, but in B2B, the physics of markets have some fundamental differences.

Nevertheless, in the spirit of studying successful messaging, what was the narrative that drove this academic work to such heights of influence?

It's our old "go bigger", story 1 approach! With a very clear from/to, old way/new way transformation to boot!

Old way/new way

The old way, from *How Brands Grow*, was "marketing managers [who] operate a bit like medieval doctors – working on anecdotal experience, impressions and myth-based explanations."

Ouch.

The new way is the science-based way backed by evidence and scientific 'laws.' The promised land, then, is *effectiveness* — brand growth through science-driven enlightenment.

Note there's a strong focus on a dominant strategy here. While the common (and deserved!) criticism of *How Brands Grow* is that it's more a case of *How Big Brands Stay Big*, it's still important to know what big brand land looks like. Even if you're a startup trying to break into the market where it doesn't initially apply, it helps to know where you're headed. Again, even if you start small, you need to know what 'big' looks like, and for our third prong of super positioning, that's being super memorable.

While *How Brands Grow* has its roots in the B2C world, these ideas are making their way into the B2B world if the number of LinkedIn posts about them is anything to go by, especially as SaaS gets increasingly commoditized and the old B2B playbooks from the 2010s run out of steam.

For our purposes in the B2B world, the most important idea we're going to steal is the idea of *brand as memory association*. This is where it comes from — *How Brands Grow* and the work of Sharp's Ehrenberg-Bass Institute (EBI).

The most important search engine

Why are memory associations a fundamental part of super positioning?

To quote the EBI and LinkedIn's B2B Institute:

> *"The most important search engine is still the one in your mind." This statement makes a profound point all marketers should internalize about buyer behavior: most purchases start not by searching Google, but by searching our memory. If you believe most buyer behavior starts with memory, it then follows that the primary job of marketing*

is not to generate clicks, but instead to generate memories.

That is to say, memory is a very big part of how people buy.

So while the B2C folks can get a little blind-men-and-the-elephant's-tail, shall we say, that still is a profound insight into the role of marketing and what creating a position in the buyer's mind actually means.

It's also neat to see how the EBI created a super position with their theory, too. Like *Play Bigger,* if we put aside what they say for a moment and look at what they do, we see:

- A strong, right-brain, story 1 pitch to put on people's radars (this is the new way!).
- In a focused, left-brain sense, they have a well-defined set of buyers that care a lot about that story (big brand CMOs).
- They've worked hard to create a strong memory association in the market, specifically that they are the source of scientific marketing truth (so please fund our research).

That's not a bad way to position brand management itself!

We won't dwell on this prong too much here, but we will pick up what being super memorable means for B2B startups in our **'own it'** strategy chapter. Part of it includes having a strong commitment to message discipling, which we'll touch on when we look at strategic messaging in **chapter 10**, too.

Synthesis

That was a speed run of the three prongs of super positioning! Go big, go narrow, or go for memory and forget the idea of segmentation altogether.

These are all useful tools to have when it comes to building a winning super position in your market. The problem is they're often presented as absolute — or near-absolute — choices. Change narrative or value positioning. Story 1 or story 2. Category or brand. Brand or positioning. And so on. You have to choose!

But do we?

If one of these approaches works for you, great. But if you're stuck, you might wonder why, and in my experience, both for marketers, sellers, and founders, getting trapped in one mode of attention can be a huge obstacle to growth.

My experience at The Unicorn, which spent the big bucks on narrative, demonstrated the limits of the story 1 alone approach.

That was all vision and narrative with no analysis, with little getting into the weeds with the product or their customers to develop their story 2 about how to manipulate the world in a new way.

The reverse can be true, too. Just as I was reviewing this chapter, I got the newsletter update of an interesting B2B SaaS I'd been following for several years.

This company had built a powerful platform, and they were doing all the 'right' positioning things, specifically loads of analysis to try and find which customers got the most value out of their tool.

But it still felt like they were going around in circles.

They'd spent *years* building a competitor to your Airtable, Notion, and Linear-style tools, but despite building it, no one much had come. Why? Well, it was all very story 2, with varying degrees of competitive differentiation but no clear, right-brain, story 1 narrative to put it on folks' radar in the first place. They were stuck too deep in the jar to see what the actual label should be. There was no meaningful hook.

I really felt for these folks because I could see the attention trap they were caught in so clearly. It was all left-brain laser beam — analyzing the features they should build (and building them), analyzing the customers that kinda sorta got it (and trying to position around them), and being stuck as to where to go next. Build and analyze, build and analyze, but no bigger-picture story 1 to connect it to.

The reality is that we need a balanced, holistic approach to positioning. The good news is that McGilchrist gives us one. Both left-brain laser beam and right-brain radar approaches can work — especially if you bring them together in the right/left/right jump. But if you get caught in one or the other and can't adapt, you get stuck, and that's what we want to avoid!

Super positioning 2x2s

I wouldn't be doing my job as a consultant without giving you a couple of 2x2s to ponder.

We've discussed our fundamental *modes* of attention — the right-brain radar and left-brain laser beam, and how they apply to traditional narrative and positioning approaches, and ultimately result in a brand memory, the third prong in the super positioning system.

They're the positioning tools for how, as I like to repeat, you get known, chosen, and remembered.

But — and this is my spin on the theory — we can also think about the *quantity* of attention we pay to things in the world as well.

For example, the quantity of attention a prospect might pay to the CRM they're considering purchasing for their sales org with 1000 reps is several orders of magnitude bigger than the drink they opt for while at the gas station on the way home.

Therefore, if we're going to map out attention, we need to consider both the *mode* and *quantity* of attention prospects use. That gives us this handy 2x2 of attention.

The Attention 2x2

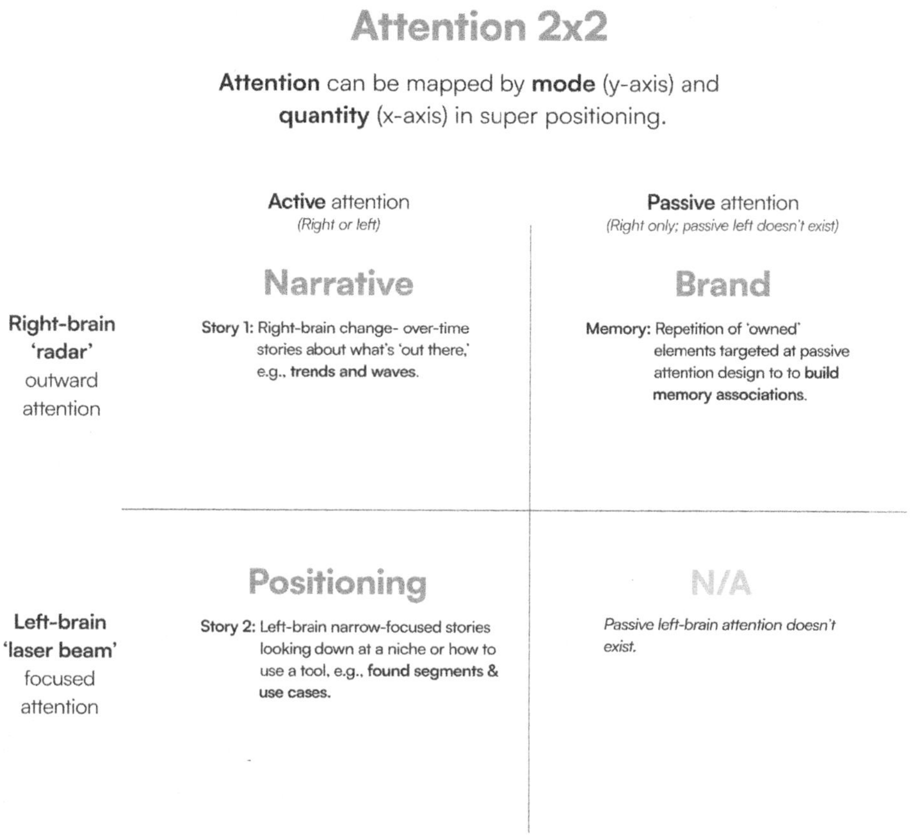

Here we see our three prongs as practical attention/technique matches, which gives us:

1. **Change-over-time narratives** for active right-brain attention.
2. **Niche-focused positioning** for active left-brain attention.
3. **Brand building** for passive right-brain attention.
4. There's no fourth kind of attention because, for our purposes, passive left-brain attention isn't really a thing.

From here, we can derive our four super positioning strategies

—**wave riders**, **problem finders**, **brand builders**, and **combiners**.

The Super Positioning 2x2

Next, if we layer on what you actually *do* to enact these strategies, we get: *ride it, find it, own it*, or *combine it* (with apologies to Daft Punk), which gives us this 2x2:

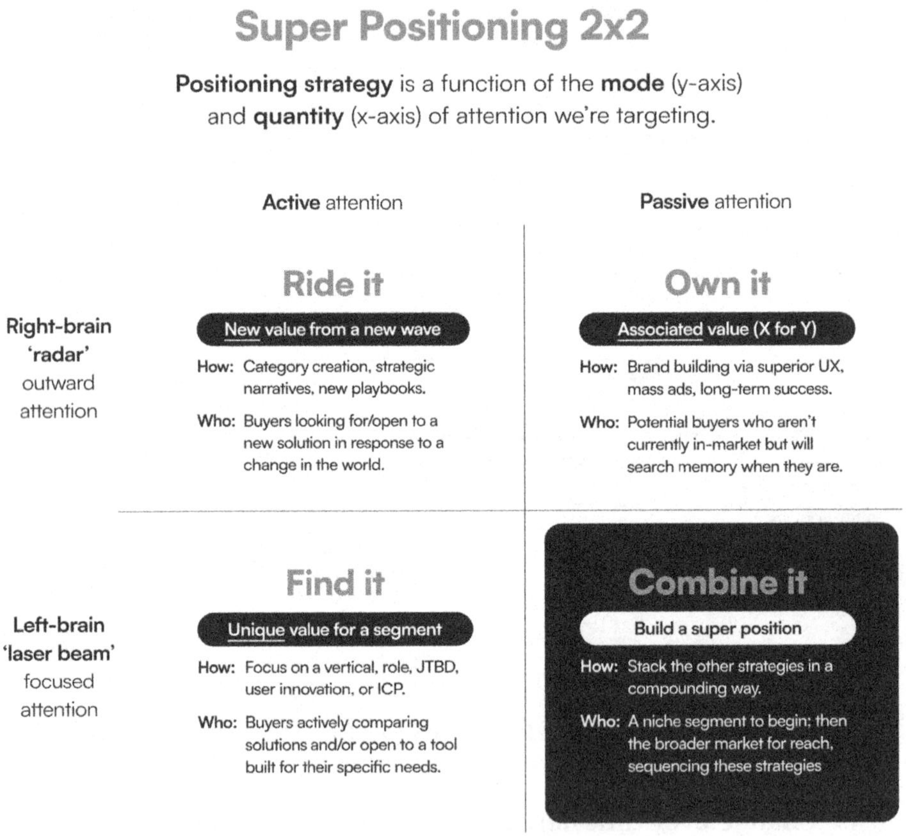

If this is all a bit much, don't stress, because we're going to unpack these four strategies, with plenty of examples, right after we dig into your strategic specific, narrative, and messaging, which is what we'll explore next.

PART III

POSITIONING EXERCISES

8

Exercise 1: Strategic specifics

The best and hardest thing about positioning is that it forces you to be specific.

Being specific creates a hook, and hooks are what get lodged in the mind of the buyer.

This means being specific about your positioning *strategy*, not just your words.

GTM folks get confused by this all the time. They see big brands using broad language and think, "Aha! They don't get it! Their broad headline is *meaningless!*" Sometimes, that's true. But big brands are specific about what they're trying to do: reach as many people as possible. They're not trying to find a niche. They're not trying to start up. They're well beyond that. And that's a legitimate and specific positioning strategy!

Startups, on the other hand, don't have that luxury. They need to break into the market, and *usually* that requires being very specific about the niche you're targeting.

(Unless you're Rippling, that is, and you're building a multi-product compounding startup, zigging where everyone else is zagging with a unique strategy!)

In this chapter, I'm going to give you a crash course on specifics in a way that both summarizes the book and gets you thinking about your positioning.

Fair warning: I'm going to throw a lot of things at you to get you thinking. I'm not trying to overwhelm you, I'm just fishing for what's unique about *you.* So in the spirit of specificity, feel free to be specific about the questions you answer and the ones you skip, too.

Running team workshops

I refer to the next three chapters as positioning exercises because they're hands-on, but you can run them as workshops with your team, too. They're designed to help you take the ideas from this book and turn them into concrete reality either individually or collectively.

To get started, we're going to tease out your *strategic specifics.*

My hope is that the next three chapters help you:

1. Flesh out the strategic specifics of your existing positioning and identify where the gaps are.
2. Create team alignment with a strategic narrative.
3. Audit and update your strategic customer-facing messaging.

That's a significant amount of work, but it's only by doing the work others can't or won't that you can find opportunities they haven't and never will.

I'm going to imagine you're doing this solo, with me asking the questions to get you thinking. However, if you're quarterbacking a more substantial positioning project with your team, you can play my role and do the question asking with your team providing answers instead.

Also, if you are doing this with your team — and positioning is a team sport, so I highly recommend you try it — it's worth creating a bit more structure in your workshops, which I won't be covering here. The good news is, however, that Google, ChatGPT, or your LLM of choice will have plenty of tips on running effective workshops, and you can make the final workshop process as formal or informal as you like. (Feel free to feed your AI this chapter and your context and get it to spit out a workshop format that's right for you!)

Grab a pen and paper and get ready to answer some questions as specifically as you can!

Being specific

Specificity (try saying that five times fast) is the name of the game here, but specific about *what?* And in *what way?*

There are specifics about your product or company that might be unique or interesting to you, but have no broader strategic implications as far as customers are concerned. Think: the programming language you used. The obscure feature or integration you built for that one customer. Even things like your mission, culture, or values might be unique, interesting, and specific to you, but they're probably not reasons for customers to buy you.

Instead, we're interested in *strategic* specifics.

Those specifics are, like just everything in this book, products of the **three prongs** of super positioning, which are:

- **Narrative:** A change in the world you've identified, like HubSpot and "inbound marketing."
- **Positioning:** A unique insight you've found, perhaps in the weird way customers used your early product, like Loom and short video sharing.
- **Brand:** A memory association you can build in the market, like Squarespace advertising on every podcast, ever, so when we hear "Squarespace" we think "websites."

They're a good start, but we want to build your positioning around your most *powerful* ideas. This isn't a tick-box exercise.

So instead, I want you to think about what makes your product or startup:

1. Super relevant.
2. Super useful.
3. Super memorable.

Powerful ideas drive powerful (super) positioning, and I'll ask you a bunch of questions to flesh out your powerful ideas below. If you get stuck or don't think your current ideas are very powerful, that's totally normal. That's why the strategy chapters exist — to help you identify and build these powerful ideas into your positioning.

But what makes a positioning idea powerful? What kind of information makes people take action?

Specifics.

There's a murderer on the loose!

Specifics create relevance. If the specifics are relevant to me, I care. If they're not, I don't.

Consider this not-entirely-serious example:

- *"There's a bad person in your city."* Ok, who cares?
- *"There's a* ***murderer*** *on the loose in your* ***neighborhood****."* Ok, now I'm scared!

Both these phrases describe the same thing. The latter, however, obviously does a better job of describing both the magnitude and proximity of the issue. The closer and more specific the danger becomes, the more I'm likely to take action!

This is a silly example, but it makes a simple point: specifics determine relevance, and relevance determines action. Things that are relevant to us are things that *feel close to us*. The more something is close to us — our location, our needs, our context, our physical safety (!) — the more we're inclined to act on it.

Rather than a feeling of danger being close by, we're obviously trying to create a feeling of *value* that's close at hand. And it is a *feeling*. We want our buyers to feel like it's a no-brainer to reach out and grab the value we're putting as close to them as possible.

Therefore (if you'll forgive the LinkedIn-worthy segue), in startups, the art of positioning is to take a product or idea and make it as *specific* and *close* to some group of people who will pay money for it.

That group of people could be very small or very large; it's not about the size of the market. It's simply about being the 'close' option in their mind.

Three steps to strategic specifics

Let's run through our super positioning prongs (➡️⬇️🔁) and see what strategic specifics emerge. These are the positioning tools we ultimately use to help you get **known**, **chosen**, and **remembered**.

1. Narrative specifics — what's super relevant?

What strategic specifics do you have that could work on the right-brain radar of narrative and change stories?

To be super relevant, you usually have to tap into a change that's happening in the world, one that's independent of you and your company. (That's why **'ride it'** — riding a wave — is the corresponding strategy here.)

Let's think about what is or could be super relevant for your customers in terms of **technology**, **tools**, and **transformation**.

Technology

- What's super relevant in terms of fundamental technology that's washing over your customers and their industry? (AI, obvs, but in what way?)
- How do you specifically bring this **general technology** closer to their actual needs?
- What super relevant **second-order effects** are occurring due to this technology? (E.g. It's now possible to automate X; or an old category like voice recognition or note-taking is new again.)
- Or, have you **created new general technology** that you're putting in customers' hands for the first time? (E.g., you're a new AI lab.) What is it? How do you describe it? What makes it so relevant?

Tools

- Have you created a **new tool or product** that's super relevant for a specific set of customers? (E.g., Superhuman for people who live in email.)
- Have you innovated on the **application layer** to get closer to how your customers actually work — and perhaps closer to their expectations for modern software, too? (E.g., Linear vs. Jira.)
- Is there some specific **outcome** you've proven your customers achieve with your tool thanks to some specific capability that's unique to you? (Think highly specific results in very niche industries. E.g., you're the *only* solution for some obscure manufacturing process, or a unique solution in some highly regulated industry.)

Transformation

- Is there an super relevant **transformation** that's sweeping an industry? That could be regulatory changes, new ways of working (agile, back in the day, or Slack pitching team chat as an email replacement), new technology-driven social behavior (social networking in the '00s, messaging in the '10s, creators in the '20s), changes in the competitive environment, and so on.
- Or, do you have a **vision for your industry** you think is super relevant? We're all in the digital transformation business, but what does that look like in your industry, in whatever wave of change you find yourself? ("Businesses should build their own tools," per Notion, for example.)
- Can you name an super relevant **new from/to** transformation? The old way is _____, the new way is _____. ("Inbound marketing" replacing "interruptive marketing," for example)
- Is there an **emerging role** your product serves you can get close to and champion? ("Design systems manager," "Customer success," or "GTM engineering," for Figma, Gainsight, and Clay, respectively.)
- Are you trying to 'create a category'? (It's ok, I won't bite!) Do you have a new **concept** or **playbook** about how work is done in

the AI age (for example) that you can put in your customer's hands? ("Experience management" in Qualtrics' case.)

Your answers here form the basis of what I call your 'story 1' change story.

(As a refresher: the two modes of attention — broad right-brain and narrow left-brain — mean there are two broad *types* of stories we're working with that we ultimately want to bring together in one narrative.)

It's possible that your innovation is so revolutionary (or specific) that it *is* the change you're trying to put on folks' radars — and if so, awesome — but try and think about trends and changes that are bigger than and separate from your product here.

In terms of examples, we'll explore what these change stories can look like in the coming chapters, but for now, we can think about these **story 1 strategic specifics** as the **drivers** of your positioning.

The x-axis of positioning

These are the bigger themes you might talk about on social media, on podcasts, in conference talks, and so on, to drive interest in your product and drive people to your website.

This is the **x-axis** of communication that's fundamentally about what you're putting on folks' radar. This is the initial creative act that's driven by your vision — the wave you saw, the product (or playbook) you built, and the initial value you proposed.

(And if you need more inspiration, jump to the **'ride it'** strategy chapter!)

Your turn

- Which approach above — tech, tool, or transformation — best matches what you're currently doing?
- Write down your 'strategic specifics' for this section.
- Give it a **specificity score** of 1-5. A score of 1 is "just a vague thought" and 5 is "customer validated and killing it in the marketplace."
- How do you rate your story 1 specifics?
- How would you improve this part of your positioning? What's

missing? What needs clarifying?

2. Positioning specifics — what's super useful?

In super positioning, we put something on folks' radar (➡️), then they switch modes of attention to analyze the competitive alternatives (⬇️).

For example, I might hear that AI makes it possible to summarize meeting notes. Maybe I hear that through some piece of marketing or social media chatter, or maybe it's just that all my colleagues are using AI meeting note takers. They might give recommendations, but I want to make up my own mind.

Therefore, I'm going to go from broad 'radar' awareness to a narrow-focus comparison of alternatives.

From the startup's point of view, this means you've successfully set the context for why they should choose *something*, but now it's time to dive into why should they choose *you*. Ideally, that's because you appear to be super useful to them.

You need to be that good to overcome the 'pain of change.' For people to switch tools, to use the old Tony Robbins quote, the pain of staying the same has to be greater than the pain of change. And that pain is driven by the insane usefulness they would otherwise be missing out on.

This is where we get narrow and focus on what's different relative to competitive alternatives. Not just peer competitors, either, but any prominent alternatives that help people solve their 'job to be done' (that might mean manual processes, spreadsheets, outdated tools, etc).

First, let's get specific about the basics:

- **Who — customers:** Who are the people you're targeting? Is it anyone with a specific need, or a particular ICP (ideal customer profile)? What did you see that made you choose them? Which segment is growing the fastest or is proving to be the most profitable? What broader industry, role, or community are you focused on?
- **What — use case/job to be done:** What's the particular job (or jobs) to be done you solve? (We'll discuss JTBD through the lens of our two modes of attention later in our **strategy**

chapters.)

- **When — triggers:** What happens in your customer's life that triggers them to consider your product? In B2C, these are called "Category Entry Points" (CEP), i.e. triggers that cause a buyer to 'enter' your category, which can be as simple as needing to refill your car and having to choose a gas station. In B2B, you might be tipped off to these CEPs through buying signals like someone changing role, growing the team, raising funds, or some other business event.
- **How — alternatives:** How *do* your customers currently complete a given job-to-be-done? What alternatives do they use? What do they compare your solution to?
- **Why — difference:** If they adopt your technology, tool, or transformation, why do they choose you? What's the *magnitude* in outcomes relative to the current way that drives their choice?

That magnitude of outcome is the key — you can have all the big ideas about change in the world and do all the zeroing in on a certain customer, but at the end of the day, the juice has to be worth the squeeze. Your product has to be super useful to the user and business as a customer.

So what's the juice for your specific customer?

If you can nail that, you can come away with some very clear messaging along the lines of:

"[Person in role], replace [alternative] and achieve [magnitude of outcome] in [job-to-be-done]."

Maybe that's **transactional lawyers** replacing **manual processes** with your new **AI-powered legal tech SaaS** that helps them **draft and review contracts** up to **10x faster**. (That example is inspired by Spellbook, as we'll explore further in our **'ride it'** chapter.)

That's both very specific and super useful! In our **'find it'** strategy chapter, we'll look at how you can develop this very specific kind of positioning, especially in terms of finding the right ICP for your product.

This is your 'story 2' competitive positioning story that matches the analytical, narrow-focus attention of customers figuring out why they should choose you.

How you get to that level of strategic specificity can vary. For example, you might focus on:

- **Wedge:** You're focused on a very specific wedge opportunity in the market — being ultra niche about a specific role, a specific task, or even a specific price point to create that insane usefulness (or accessibility).
- **Attribute:** You might go all-in on a specific attribute (collaboration, automation, intelligence, etc.) that you think drives insane usefulness for your ICP.
- **Proximity:** You might just be better at getting closer to the customer, both in terms of understanding their jobs-to-be-done, what they value, or how they buy, and messaging that back to them. You're the super useful glove that fits perfectly.
- **Weirdness:** You may have spotted a 'user innovator' doing something weird — and very interesting — with your product. A surprising amount of startup success is found in zeroing in on the specific — and super useful — thing some customers do with your product, which may not be what you'd thought they'd do!
- **Sameness:** Everyone in B2B bemoans the 'sea of sameness,' but if you're competing with an incumbent, you inevitably need to 'compete to be the same' (and not just 'compete to be unique') to some degree. That can be both necessary and highly strategic, so feel free to include it as a strategic positioning specific!

The y-axis of positioning

This is the **y-axis** of comparative difference. In the world of B2B positioning, rather than trying to guess what *might* be super useful, these strategic specifics are often found by **looking down** at who is already getting insane value out of your tool or platform and then doubling down on them.

Note that we've shifted from the strategic specifics of your **x-axis outward vision** (the opportunity and the initial product) to your **y-axis customer vision** — analyzing who's choosing you, why, and what outcomes they're achieving. This is the same two modes of attention at play — looking out and looking down — as those we're trying to target.

Your turn

- Who finds you super useful? Do you have an 'ideal customer profile' (ICP)? Is that speculation (you're still exploring your market) or validated by studying your customer base (you've found a segment to exploit)?
- What job-to-be-done do they choose you for?
- What are the common alternatives? How do you know?
- What triggers them to enter your category? Again, how do you know?
- Write down a classic positioning statement along the lines of "We're an [category] for [ICP], who choose us for [job-to-be-done] because unlike with [alternatives] they achieve [outcome]."
- Give it a **specificity score** of 1-5. Again, think of a score of 1 as "vague and hand-wavy" and 5 as "proven and validated through detailed analysis and customer contact."
- How do you rate your story 2 specifics?
- How would you improve this part of your positioning?

(Note that positioning statements are usually pretty lame because they describe usually poor positioning as it is; they don't help you find a better position. But here they're useful as a diagnostic tool to see just how specific you can get!)

3. Brand specifics — what's super memorable?

Brand is the poorly understood, red-headed step-child of B2B sales and marketing. And that's a shame, because the universal destination of all positioning work is a memory association you build in the mind of the buyer (and market at large). It turns out that's the modern definition of brand that the B2C folks figured out long ago!

If our first two steps in this chapter were "what's new & relevant" () and "what's different & useful" (), this section is about "what's memorable" ().

You might think, "Isn't the memory we want to build what

we've already talked about, like the big change or our specific differentiation or usefulness?"

That's true to a point — HubSpot was famous in part for its "inbound marketing" concept. Gainsight was famous for growing the Customer Success category.

But here we're talking about the residual, low-resolution takeaway. Hear Squarespace, think websites. Hear ChatGPT, think AI. Hear Figma, think design.

Very few companies can own an association with their entire category, but startup success ultimately demands *some* kind of memory association between your product and the need you serve. You can either just let that happen (or not happen), or you can be deliberate about how you'll be — and *remain* — super memorable in your specific market.

I say remain because memories *decay*. This isn't just about making a big splash assuming everyone will be so impressed with your greatness you'll be instantly burned into their memory forever more, it's about being realistic about the sustained effort it takes to both build and maintain that key memory association in your market.

The good news is that when the market knows you and remembers you, *everything* gets easier. You become the first solution that pops into someone's mind *before* they search Google or ask ChatGPT. The familiarity you've created makes your marketing and outbound sales vastly more effective when you reach out. And other people will happily recommend you, even if they've never used you. They just know that if you need a website, maybe you should try Squarespace. (Or Webflow, if you're in SaaS.)

That's the power of brand. It's vastly more than your logo or any of the 'high-res' specifics you associate with it; it's about that super memorable hook that's buried in your memory as a name/need association. Hear X, think Y.

From niche to reach

Being super memorable in that sense matters, but the big question is, again, for whom?

In product marketing in particular, we're so used to being

specific on the focused, story 2, competitive positioning side that we can forget that, ultimately, building a brand is about switching from *niche* to *reach*. Reach means selling more stuff to more people, which is how you grow.

Founders sometimes intuitively know that's what they want to do, and this tension emerges between founders pulling towards reach and GTM folks pulling towards niche, at least to get started.

I've seen this first hand, where the founder in an all-hands meeting was trying to fire up the team about conquering the market, while their CRO was trying to very politely but very pointedly suggest they focus on a single niche first. They weren't on the same page in terms of their niche or reach specifics.

To get back on the same page, you need to get folks agreeing on the sequence you want to take. Everyone wants the big outcome in the end — with the maximum amount of reach that entails — but the question is how you get there.

The classic positioning advice about getting narrow, finding a niche, and focusing just on your best buyers can be great to get started. When there's a whole sea of possibilities (including all the ones I'm throwing at you here), it can certainly simplify things to just focus on ONE niche and ONE ICP to get the ball rolling.

However, if you are post-product/market fit, then finding a niche becomes less relevant — and maybe even detrimental! — as you should be expanding your brand reach and building those memory connections at scale.

Know which game you're playing.

Reach where, specifically?

Being specific about brand and reach means being specific about where you're trying to expand your reach. You might reach up or down market, reach into adjacent niches, or just flat out reach more people using B2C-style brand marketing. Either way, it's worth being strategically specific about that, too!

For more on owning associated value with your brand, check out the **'own it'** chapter, or press ahead and we'll work out your strategic specifics score.

Your turn

- What's the super memorable name/need association you're building in your marketplace?
- How memorable are the specifics in your story 1 or story 2 pitch?
- When people hear [your brand], they'll think... what? What's the association?
- Write down the most **focused** version of your brand hook — what would it say?
- Write down the most **expansive** version of your brand hook that you aspire to — what would that be?
- Now, write down where you currently sit on the niche-to-reach spectrum. Are you trying to get a foothold to get started, or are you trying to expand your reach?
- What's the next move on building your brand? Is it getting more niche to build more momentum, or is it expanding your reach outwards using the momentum you've built?
- Give your brand positioning strategy a **specificity score** of 1-5. A score of 1 is "we're dreaming," and 5 is "random people associate our brand with our chosen need."
- How would you improve this part of your positioning?

Your strategic specifics score

Let's recap. Write down your:

- **Narrative strategic specifics** — the things that make you super relevant — and score (1-5).
- **Positioning strategic specifics** — the things that make you super useful — and score (1-5).
- **Brand strategic specifics** — the things that make you super memorable — and score (1-5).

Now tally up your scores.

If you scored **15** in total — awesome! You are very strategically specific and you've got a lot of great ingredients for your narrative, which we'll look at next.

If you scored **10** in total (give or take), you've got some serious strengths to work with! You can consider how you can double down on them, or complement them with one of the prongs where you're weak.

If you scored **5** or less, don't worry, you're just getting started! You're in the right place — everything in this book is designed to help you get as specific as possible about each positioning prong.

Your positioning personality

Your strategic specifics score is just a fun way of positioning *your own* positioning in your mind, if I can put it that way, by quantifying where you are and where you need to be.

Here's another exercise that can be fun to think about and useful to discuss, too.

If we put **vision** and **validation** on separate axes, and add a brand layer, we get the Super Positioning Personality 2x2:

Positioning Personality 2x2

Vision is mapped on the x-axis and **validation** on the y-axis in super positioning.

Validation
(Story 2)

Operator

Story 2: Strong analytical take on what's working.

Story 1: May lack (or completely dismiss) the overarching narrative.

Super Positioner

Strong story 1 & story 2: Both stories are working together in a unified way.

This has created a strong hook in the mind of your buyer and market.

Vision
(Story 1)

Early/Drifter

Weak story 1 & story 2. Positioning purgatory. Either too early to have developed a meaningful vision or set of customers to analyze, or have been drifting through the market. Time for decisions!

Visionary

Story 1: Strong focus on the big idea, the big outcome, and what 'could be'.

Story 2: May struggle to focus down on what is; may expect world to come to them.

Brand ambition

Your orientation to your market defined by stage, scale, resources, & ambition

Niche		**Reach**
Sales-led	→	Marketing-led
"Segment & sell"		"Cast & catch"

As you can see, this gives us:

- **Super positioners:** You're scoring close to 15 and have nailed your story 1 and story 2 positioning across the vision *and* validation axes.
- **Visionaries:** You may be strong on story 1 vision but weak on down-with-your-customers validation.
- **Operators:** Or you may be excellent at understanding the story 2 customer-specifics, but struggle to bring a new vision to them.
- **Early/drifters:** You're too early, or have been drifting in the market without much deliberate positioning, and are stuck in positioning purgatory.

Add on your brand ambition, and, for visionaries and operators, you might be a:

- **Visionary with Niche Ambition:** You have big ideas, but you're happy to test them in a small niche.
- **Visionary with Reach Ambition:** You want to go all out and 'create a category.'
- **Operator with Niche Ambition:** You're happy to focus on a niche and take a very methodical approach to studying the needs of your ICP and gaps in the market.
- **Operator with Reach Ambition:** You're about building sophisticated GTM plans to reach the mass market.

I know we all *love* to see ourselves in (often fairly dubious) personality quizzes, so don't take this too seriously. Maybe it highlights some strengths and weaknesses, but the point isn't to box you in. I want to show you the possibilities you may not naturally consider — maybe you should be more of a visionary! Maybe you should just focus on a niche! Keep those things you might otherwise dismiss in mind as we go through the rest of the book — you might be surprised what you're open to by the time we're done.

Compounding & the super positioning whole

We now (hopefully!) have a bunch of strategic specifics that we can work into your broader positioning. Awesome! Good on you for making it through this first positioning exercise. Your collection of specifics probably feels a bit loose and disorganized at this point, and that's totally fine. In the next chapter, we'll get more organized with your strategic narrative, and then we'll really boil things down for your strategic, customer-facing messaging in the chapter after that.

You might also have a sense of the kind of story you *don't* have. And, again, that's fine — this is also a diagnostic exercise to see where you're strong and where you could use help. There are plenty of insights and examples to come to help you fill those gaps.

But where are we going with all of this?

There's something bigger you need to keep in mind as the super positioning quarterback for your startup, too.

In super positioning, however, we're not trying to land on The One Decisive Specific that will make the rest of the market cower before us; we're trying to find a collection of integrated specifics that create something greater than the sum of its parts. That's the new 'whole' we're trying to get to.

This is the difference between silver bullet thinking and compounding thinking.

Everyone loves to try and zero in on THIS tactic or THIS approach or THIS feature that will take your growth to the next level, but the reality is always AND instead. And, and, and. Relevant AND useful AND memorable, for example. It's bringing those most powerful ideas together and giving them time to marinate that helps you eventually build a company that's just so recognizably *that company.*

Notion

Take a company like Notion, for example.

We'll discuss this more later in the book, but they had a strong story 1 vision (user-built software and a highly novel product), they developed a clear story 2 pitch out of necessity (an initial focus on wikis, docs, and project management), and they wrapped it up in a very distinctive brand (including their initial black-and-white visual assets).

And, and, and. That's the compounding approach. Bringing these things together achieved the final brand goal of creating a memory association between a wide variety of productivity tasks and their product, giving them a lot of reach.

That's what we're aiming for — bringing all three of these prongs ( ) together in a novel, unified, and compelling way.

Drift

Likewise, as we'll touch on Drift again in our **ride it** chapter, but Drift took an existing technology (web chat) in a competitive space and built a super position through compounding choices. They very deliberately rode a wave (the rise of messaging — their story 1), found a very specific differentiation opportunity in a specific industry (use chat, not forms, for sales and marketing — their story 2), and owned it with a very distinct brand (remember the lightening bolt?) along with an owned concept ('conversational messaging'). That made them, for a time, one of the most recognizable brands in the industry — a brand that was, inexplicably, almost wiped from the face of the earth post-acquisition!

Strategic specifics to strategic narrative

Compounding choices are the essence of strategy, and ideally, that 'whole' you build engages as much of your target customer's brain as possible. That's why super positioning is about engaging the right-brain radar, left-brain laser beam, *and* brand memory all in one unified whole.

But how do you get the team aligned around that, on the one hand, and have something to put in front of actual customers to get their reaction, on the other?

You need a strategic narrative, and that's what we'll look at next.

9

Exercise 2: Strategic narrative

Creating a strategic sales narrative is the focus of our second positioning exercise and, in some ways, this entire book! It's one of the most crucial things you can do to work out your super position simply because it captures the life-changing magic of Writing. Things. Down.

The sooner you can get your ideas out of your head and onto the proverbial page, the easier it will be to stretch them, test them, and ultimately align your team around them.

Therefore, in this chapter, we're going to get hands-on and start fleshing out a real, live strategic sales narrative that you can test with prospects.

Memos, narratives, and pitches

First, let's tease out what we're aiming for here. Why, for instance, are we focused on a strategic *narrative* here and not an internal strategic memo or a final sales deck for your sellers? (And what's the difference, anyway?)

When it comes to positioning and strategy, there's a spectrum from internal to external communication:

- **Internal:** A **memo** or document that can take a variety of forms (prose, bullets, slides, etc.) that exists to *describe* the strategy internally.
- **Internal *and* external:** A structured, repeatable **story** that exists to *sell* the strategy internally and externally to employees, investors, and relevant customers.
- **External:** A final, customer-centric sales **pitch** that's designed specifically to meet the prospect where their attention is at.

For our purposes, the job we're giving our narrative is articulating and selling the big ideas that animate your startup using our R/L/R pattern of context, specifics, whole.

Later in the chapter we'll segue from the higher, more strategic level to something more akin to a sales pitch.

Before we do either, though, let's take a brief look at one of the greatest internal strategy memos of all time.

"We don't sell saddles here"

I'm going to walk you through a structured strategic narrative in a moment, but I consider my outline the *floor* for what a narrative can be, not the ceiling. You are — and absolutely should — run with it and make it yours. What we're discussing here is just scaffolding.

To give you an idea of what the *ceiling* for internal strategic communication looks like, however, it's worth touching on Slack founder Stewart Butterfield's classic "We Don't Sell Saddles Here" memo from 2013 (!), sent out to the team before Slack launched.

Butterfield's aim with his inimitable strategic story was to focus the team on the opportunity ahead of them — both on the product *and* market side of the product/market fit equation. Specifically:

> *Our position is different than the one many new companies find themselves in: we are not battling it out in a large, well-defined market with clear incumbents (which is why we can't get away with "Other group chat products are poisonous. Slack is toasted."). Despite the fact that there are a handful of direct competitors and a muddled history of superficially similar tools, we are setting out to define a new market. And that means we can't limit ourselves to tweaking the product; we need to tweak the market too.*

Tweak the market, indeed. Slack, of course, went on to build a super position (and poke the Microsoft bear), and that super position was evidenced by behavior change — lots of teams adopted group chat! Butterfield continues:

> ***Sell the innovation, not the product*** *The best — maybe the only? — real, direct measure of "innovation" is change in human behaviour. In fact, it is useful to take this way of thinking as definitional: innovation is the sum of change across the whole system, not a thing which causes a change in how people behave. No small innovation ever caused a large shift in how people spend their time and no large one has ever failed to do so.*

Butterfield wanted to rally his team to sell the behavior and its outcomes, not "chat" per se:

> *That's why what we're selling is organizational transformation. The software just happens to be the part we're able to build & ship (and the means for us to get our cut).*
>
> *We're selling a reduction in information overload, relief from stress, and a new ability to extract the enormous value of hitherto useless corporate archives. We're selling better organizations, better teams. That's a good thing for people to buy and it is a much better thing for us to sell in the long run. We will be successful to the extent that we create better teams.*

Or, to use his metaphor of the hypothetical Acme Saddle Company, "They could just sell saddles [...] Or, they could sell *horseback riding.*" (Emphasis in the original.)

That is, it was about the big idea — not just saddles, not just chat, not just features, and definitely not price. That's what super positioning is about.

While Butterfield's approach was an internal memo, and we're going to look at something more akin to a sales narrative, the gist is the same. We're primarily interested in the big, story 1 ideas that you're putting on folks' radars.

Consider how different those juicy outcomes Butterfield mentions are — less information overload, more value from past communication, better teams — from the chat features of Slack itself. Yes, they put great care into their UX and design (when they launched in 2014, it was something of a revelation!), but it was their story 1 focus that helped them *transcend* group chat as a category. Teams didn't run on chat; they ran on *Slack.*

That's the kind of big idea — and big outcomes — we want to capture here in your strategic narrative.

While Butterfield gave some suggestions for *how* the team would sell and promote Slack, he was setting parameters more than dictating a final deck or story. That's the difference between an internal memo and the strategic narrative we're looking at here. The memo can absolutely work, but we want something more structured.

Story 1 space

While we're on Slack, it's worth noting how Butterfield, as the founder, gave the team — and the tech press — room to stretch their thinking about what Slack could be in a story 1 sense beyond mere chat.

While Slack eventually became completely synonymous with chat, when it launched in 2014, it was a weird new thing. They had grand strategic ambitions of being the hub for all your business software. They dreamed of becoming the next Microsoft (ouch). And that dream of being that central hub never really died. In 2025, Salesforce attempted to relaunch Slack as an "Agentic OS" in the AI era and the "conversational interface for Salesforce." A new narrative for a new age. Whether it sticks or not remains to be seen, but that story 1 space creates room for aspirations and experimentation in a way that a pure feature focus rarely does.

The strategy in your strategic narrative

What makes a narrative "strategic," though?

The point of super positioning is to transcend your category with an idea that's bigger than your product alone.

That requires making clear choices about how you'll define and communicate the position you want to build, as Butterfield did above.

The strategic narrative, then, is where we capture those choices and sell the big idea using our R/L/R framework.

A strategic narrative, therefore, does two things:

- **Your choices:** It captures the choices you're making as an organization — the context you see, the market you serve, the way you serve your customers, and the outcome you deliver for them.
- **Their choices:** It captures the choices your *audience* is facing as they confront some change in the world or a new way to manipulate the world.

As a leadership tool, then, it's a convenient way of conveying your story (and your choices) to your own employees, your investors, and perhaps your broader market.

As a *sales* tool, however, the full strategic narrative is useful *only to the extent that the audience is grappling with the choices you capture.*

This is what separates your strategic narrative from a sales pitch at the other end of the spectrum.

The 4-ply pitch

If I'm going door to door pitching, let's say, the latest in high-quality toilet paper to small businesses, I might have a tight sales pitch for my product. But there's nothing especially *strategic* about it. I'm not pitching executives about the strategic relevance of adopting my soft, absorbent, velvety, 4-ply.

Contrast that with B2B enterprise sales where, at the high end, companies might be spending millions of dollars per year on contracts for your platform and considerable support services. In that case, you'd better believe there needs to be some strategic imperative for an enterprise to bet heavily on your company or startup.

As you can imagine, the latter case is story 1-driven (the change or new innovation that drives action), while the former toilet paper case is story 2-driven (why are you different from my current scratchy-but-cheap toilet paper?).

Escaping the laser beam

Most B2B sales conversations happen at the story 2 end, where the discussion quickly turns to "How are you different?" and "Can you solve my very specific problem?"

For example, Daniel Zarick, founder of sales rooms and onboarding SaaS company Arrows, posted on Twitter/X about how his prospects were showing up in sales calls:

> *Our sales calls these days are often super specific questions. "I watched a bunch of your YouTube videos... can I do X and Y via the integration?" We still do a structured sales demo, but it's almost like solutions engineering half the time.*

The fact that buyers do much of their own research and only show up very late in the sales process is a well-documented B2B phenomenon.

If you, as a founder, seller, or marketer, are constantly engaging with the prospect's left-brain laser beam, it can be tempting to think that feature bingo or sales engineering is what sales and marketing (and indeed positioning itself!) are all about.

But what's the big idea that brought the prospect to you in the first place?

What's the big idea that you can put on their radar in that crucial first meeting (*along with* your superior feature set) such that it raises their aspirations for what they can achieve in their role at work?

What's the sales judo flip you can pull to reorient the conversation around the new, high-level whole you see — the bigger outcomes, the more meaningful impact — rather than just the story 2 nitty-gritty of features and integrations?

That's what we're going to tackle in this chapter.

Pick your context

To pull this off, you have to pick your context. If you pluck a successful narrative that does a great job with the C-suite and drop it on inbound end-users, for example, it can still fail to resonate — and perhaps even annoy! — because of the story 1/story 2 mismatch.

If someone enters a sales call super late in the buying process in left-brain laser-beam mode and you hit them with "A change in the world we observed was..." they're going to be irritated at you for wasting their time. That's not very strategic.

If those end-user buyers already understand the context and just want to know *if this will be useful for me*, then pitching them on why it's relevant in the first place is obviously a mistake. It's like telling someone why Cheetos are a great snack when they've already opened the pack, and their fingers are covered in orange dust.

However!

You *can* still pull the conversation back up from a specific feature or integration to the bigger picture about what they're trying to achieve.

You *can* still probe to see if they're open to a more story 1-style conversation.

You *can* still use the story 1 driven approach with people in that mode of attention (execs and interested outbound folks, for example).

You can still *sell*.

You're not just there as a human FAQ section. You can work with both modes of attention to have the kinds of mutually-beneficial conversations you want to be having.

Early adopter stampede

Likewise, in the midst of the AI supercycle, there are hordes of early adopters who have been told by their superiors to go out and adopt something — anything! — so we can see what all this AI fuss is about.

This can negate the foundation of Andy Raskin's original strategic narrative, where he argued:

> *No matter what you're selling, your most formidable obstacle is prospects' adherence to the status quo. Your primary adversary, in other words, is a voice inside people's heads that goes, We've gotten along just fine without it, and we'll always be fine without it.*
>
> *How do you overcome that? By demonstrating that the world has changed in such a fundamental way that prospects have to change, too.*

That was true at the time — the point was to make the case for a change that was independent of your company in a mature SaaS world.

In the AI era, though, everyone is being hit over the head with the case for change. Top-down mandates are literally telling people 'the world has changed in a fundamental way, and you have to change [i.e., find an AI solution] too.'

In this case, it's again not very strategic to pitch the buyer on the change itself (they're well aware!). But it's still incredibly important to have a story 1-style case for your big idea *in light of that change* — the one that helps you transcend your category, drives your perspective, hits the second-order effects prospects need to know, and underpins your theory of their success, both personally and for their organization as a whole.

The job of a strategic narrative

While we're going to get your ideas down into a tight narrative, the strategic thing to do, however, is seed those ideas in the market so your audience can consume them *before* they come inbound to a sales call. Then, if you want customers to stick around, you have to close the loop between your big idea and the actual tangible success they achieve.

That's what we're trying to capture with our strategic narrative in super positioning — writing down your big ideas in a concise and catchy way (R), demonstrating how your product achieves that (L), and closing the loop with a new whole about the success your customer achieves (R).

Behavior change

At the heart of your narrative is behavior change.

That goes back to the origins of the strategic narrative, both as a concept and in business. Amy Zalman originally coined the term "strategic narrative" in the early 2000s in the context of international relations. In 2022, Zalman wrote that:

> *[It] has become clear to me that strategic foresight projects [i.e., vision-based projects] are almost always missing a critical element: building a strategic narrative that models new behaviors to help spread and scale your preferred future. Without a new story, people won't behave in new ways, and they won't rewrite conditions on the ground.*

I love that idea: 'Without a new story, people won't behave in new ways.' Behavior change — i.e., using your product and playbook, not just buying your thing — is ultimately what we're after with super positioning. It's what Stewart Butterfield observed with Slack ("The best [...] real, direct measure of "innovation" is change in human behaviour"). Plus, a focus on behavior change helps keep us grounded, especially when folks can get carried away with the strategic narrative concept.

The strategic *business* narrative

Sometimes people *can* get a little carried away with the possibilities of a strategic narrative.

For example, in March 2016, Mark Bonchek wrote an article for *Harvard Business Review* titled "How to Build a Strategic

Narrative" in which he said:

> *It's not enough any more to say "we make widgets." [...] You want a story that inspires employees, excites partners, attracts customers, and engages influencers. A story that is concise but comprehensive. Specific but with room to grow. One that defines the company's vision, communicates the strategy, and embodies the culture. [...]*

That's a lot of jobs for a story to do! In fact, that's too many jobs for a story to do.

Writing in September of the same year, Andy Raskin wrote a classic viral essay that formed the basis of the strategic narrative in tech circles, which was the very story 1-heavy approach we've looked at as the first **prong** of super positioning.

Raskin had the same broad ambitions as Bonchek — a single story that can inspire employees, attract customers, and generally act as a leadership tool — but gave it a *very* specific focus: a change story.

These days, in light of the AI era, as we touched on above, that's too narrow.

But, again, there's a deeper principle at play — and that's having a big category-transcending story 1 idea *at all.*

The big idea in Raskin's approach *was* the change story. It was based on Zuora's high-level enterprise sales pitch, which revolved around the concept of the 'subscription economy.' That was the change story — there would be winners and losers, and Zuora's features would take you to the 'promised land' as a winner on the other side of the big change. (Remember, this worked with executives because they were in right-brain story 1 mode.)

The big idea in category creation is supposedly the category, but really it's the concept you're bringing to the category.

Even sales classics like *The Challenger Sale* and *The Challenger Customer* talk about 'commercial insight' as the big idea you're trying to put on the buyer's radar, perhaps by breaking down their existing mental model and creating a new one.

And as we saw with Slack, their ability to carve out that more ambitious story 1 space helped create room for them to transcend "group chat" as a category, and opened up new ways for people to think about and succeed with the product.

Like the blind men and the elephant, everyone has their own particular take on it, but the underlying reality for these story 1 approaches is the same: it's about having some new idea that's bigger than your product that you can put on folks' right-brain radar.

Adding your twist

What does that look like in practice? This can come in a variety of flavors, and we'll explore a bunch of examples in our **strategy section**.

In terms of how you actually pitch, though, you have to think hard about what's true for you.

Understanding vs. manipulating

Snowflake, for example, when they were one of the most formidable sales operations on the planet, had what Benn Stancil humorously described as "the dumbest and most effective sales pitch I've ever heard":

> *We were told that [Snowflake] was the same as Redshift—and really, the same as Postgres—but big, fast, and stable. [...] There were no forced mentions of digital transformations, distributed systems, or, God forbid, the blockchain. They never even told us what we could use Snowflake for. We were told to just keep doing what we'd been doing, except if we did it with Snowflake, they'd host the whole thing, back it with bottomless storage, and run our queries faster than ever before.*

Here, Snowflake's product innovation was significant enough that that's all reps had to put on folks' radars. That was what was true for them. The big idea was simply that the product was bigger and faster, and switching was easy because they didn't have to do anything *other than use Snowflake*.

Contrast that with Zuora's approach: one was a new way to *understand* the world (the economy is changing); one was a new way to *manipulate* the world (or your data, in this case).

The point isn't that there's one right — the point is that there are two modes of attention and you need to find the right story 1/story 2 framing that most makes sense for your audience and market. For example:

- **Story 1:** A cybersecurity product might talk much more about the emerging threat 'out there' (AI-enabled social engineering or prompt-injection exploits, for example). As a prospect, I need to understand what should be on my radar before any solution will make sense, so more of the narrative will be spent setting that context.
- **Story 2:** A developer-oriented product might get into the specifics of technical details, APIs, and integrations — and perhaps the challenge of build vs. buy — very quickly. Any change 'out there' *apart from the innovation itself* is less important than the specifics of how it works, so we might lean more on a story 2-style pitch in that case.

In doing research for this book, I found B2B SaaS CMOs had adapted the classic strategic narrative in a variety of ways. They gave it their own twist to suit their market and their needs at the time.

Concept-level differentiation

Ryan Yackel, a CMO who has worked across several successful startups, has had repeated success with the *concept-driven differentiation* a strategic narrative can encourage.

On the *On Messaging Podcast*, Yackel described one example where, at software testing company Tricentis, the category the company operated in — and Gartner *et al.* acknowledged — was "test automation."

Tricentis had come up with a better, more holistic strategy for testing, so they called it "continuous testing" instead. That spoke to all the different kinds of testing they offer their customers.

The term itself wasn't magic. The point *wasn't* that this was a new category. The point was simply that it captured their *concept*, creating a differentiated narrative. That, in turn, helped the company *transcend* the typical test automation category comparison and shape conversations around Tricentis' strengths and their suite of testing offerings.

Again: they could have the sales conversations they wanted to have because they could get back to the story 1, concept-level differentiation.

That's often the recurring theme of strategic narrative work — a simple hook, idea, or catchy concept that helps folks transcend their category. Those memetic 'information packets with attitude' can be immensely valuable things!

Framing your narrative

Likewise, MJ Smith, CMO at CoLab, a collaboration platform in the manufacturing space that streamlines the design review process, gave the classic strategic narrative her own unique spin.

Inspired by Raskin's approach, Smith and her team spent months doing the legwork of customer research, internal alignment, and sales call analysis. They got the bulk of their narrative down without too much trouble, and that covered the outcome, features, and evidence they could deliver.

In the super positioning world, we'd say they had a good handle on their story 2 pitch, but Smith struggled with the story 1 side. What made them relevant? What was the change and what was at stake? What did winners and losers look like? What were they putting on folks' radars?

Ultimately, Smith came up with a novel framing device: Company A and Company B, "two manufacturers on different trajectories," where she could show in a highly visual way the compounding advantages of using CoLab. The advantages started with speeding up design reviews (faster reviews → better decisions → better products → stronger brand) and the disadvantages of not using CoLab (errors slip through → launches are delayed → more warranty claims → lower customer trust.)

The reason this worked for their narrative is the story 1 focus. It's about change over time *for the customer.* Smith literally puts these two paths on a diagram with the winning path curving up over time and the bad path curving down over time. Which would you rather be?

That approach neatly framed their story — they had a well-thought-through theory of customer success and the "Company A/Company B" approach became a part of their sales narrative *and* a part of their internal thinking. Smith wrote that:

> *Not only is Company A / Company B a key part of our sales narrative, it has become "a thing" internally at CoLab too. We talk about our own wins as "Company A" behavior. When a prospect is dragging their feet, we're struggling to overcome "Company B" behavior.*

Your turn

- Are you leaning more on story 1 or story 2?
- Are you presenting a new way to understand the world, or a new way to manipulate the world?
- What's the primary mode of attention for your audience? Are they executives focused on macro trends (story 1) or end-users more focused on how the tool works (story 2)?

Strategic narratives in super positioning

We've covered strategic narratives in general — where they fit between memos and pitches, why you have to give them a specific job around behavior change, and why they're a chance to start with your story 1, whatever that might be.

Now let's start thinking about your strategic narrative.

We've established that story 1 and story 2 exist as different stories, targeting right and left-brain modes of attention, but here, with the strategic narrative, we bring these two modes of attention together using the R/L/R jump.

This is based on McGilchrist's super useful insight about how our brains process information.

Remember that, even for very story 2-centric folks, you still have to put *something* on their radar in the first place, like the fact that your new product and innovation exists at all. Likewise, for story 1 folks, there's still *some* getting into the details to prove your value really is there.

This gives us a pattern we can use for *any* strategic (or sales) narrative we might be putting together.

Broadly speaking, we go from **global context → narrow specifics → new whole**. That is, we go from:

- **Right hemisphere:** What's new, what's changed, what's the fresh insight, what's 'out there' on the radar.
- **To the left hemisphere:** Applying our narrow focus to drill down to the parts and pieces.
- **Back to the right hemisphere:** Coming back up with a new, integrated whole.

This is the R/L/R jump: context → specifics → whole.

Story 1-focused narratives spend more time on the first R; story 2-focused narratives spend more time on the middle L, and for both narratives, we should always bring it back to some new world that buyers will experience on the other side.

It's thinking through that last part, in particular, that helps us shape our micro-pitch, our narrative, and ultimately the position we're trying to build in the buyer's mind. What will things look like for the buyer on the other side?

Your turn

- Does your current narrative or sales pitch map to this right/left/right, or context/specifics/whole approach?
- Where are you strong and where are you weak in terms of setting context, getting into specifics, and bringing it back to a new whole?
- Does that final step — the new, reintegrated 'whole' — reflect what you've learned from customers?
- How does your long-form content (articles, ebooks, webinars, demos, etc.) map to this approach? (Again, in a sense, a startup is just a nested set of R/L/R loops!)

Specifics and the value you propose

Before we get into the weeds of the actual slides, let's remember that part of the job of your strategic narrative is to take those "strategic specifics" we teased out in the previous chapter and roll them into a coherent value proposition, to use the marketing term.

Your value proposition is the glue between your product and market that goes to the heart of the position you're trying to build. Think of your value proposition like the power cord that connects product and market together, in a product/market fit sense. If that power cord doesn't connect... well, not much value is going to be created.

That's why your strategic narrative sits in-between your internal (company) and external (customer) worlds — we're trying to find the story where they connect.

To do that in an interesting way, we need specifics that are highly contextual to you. Without them, we get generic fluff that sounds like everyone else. That's why we spent so much time fleshing out your strategic specifics in the previous chapter — they're the raw ingredients that help us articulate your value proposition and for whom that matters. That's what buyers are ultimately going to buy, after all: the value you can start delivering. But to communicate that in an interesting way you *must* have the specific product or market insights that we can use.

So keep those strategic specifics handy, and then let's run through:

1. Framing your narrative (story 1 & story 2).
2. Concept naming.
3. Your ~10-slide strategic narrative.
4. Your tailored sales pitch.
5. How you can create further value in your pitch.

1. Framing your narrative (story 1/story 2)

Hopefully you've been through the **strategic specifics** in chapter 8 and have a list of:

- **Narrative specifics** — what's super relevant? (We covered tech, tool, and transformation.)
- **Positioning specifics** — what's super useful? (We covered who/customers, when/triggers, how/alternatives, why/difference.)
- **Brand specifics** — what's super memorable? (What's the

memory association you're trying to build in the market?)

The big question, then, is whether you are fundamentally pitching:

- **Story 1 — a new way to *understand* the world**, with a right-brain, story 1 change pitch? That was the idea behind Zuroa's pitch — the world had changed, the subscription economy was here, and they had the product to help you play this new game.
- **Story 2 — a new way to *manipulate* the world**, with a left-brain, tool-focused, story 2 positioning pitch, like Snowflake? Just want a better [your category of tool]? That's us. Nothing has changed; this is just better technology for managing your data in the exact same way as you were doing it before.

Zuroa had a market insight; Snowflake had a product insight. What's yours that helps you frame your narrative?

Remember, the best ideas — and the best framing devices — are often simple and catchy. CoLab had "Company A/Company B"; Tricentis had continuous testing vs. test automation; and the hooks we looked at earlier in **chapter 3** could be expressed in just a few words. That included classics like "Use cloud software not on-prem" (Salesforce), replacing 'interruptive' marketing with 'inbound' marketing (HubSpot), or even just "payments for developers" (Stripe).

So, first up: What's the simplest way you can frame your narrative?

2. Concept naming

While the core of your narrative universe is your hook, you don't have to stop there.

Let's try fleshing out your world a little bit by naming concepts where you have a compelling insight into your buyer's world — a pain, a solution, a playbook — something that's unique to your point of view.

If you've got your strategic specifics ready to go, great; if not, or if you need some more ideas, here are some more suggestions.

(Again, get the pen and paper out and get writing!)

Context

- **Problem:** What's your name for the burning pain or impending doom you're tackling? Is there a specific trap, challenge, or hidden or missed opportunity you can name?
- **Winners/losers:** Can you frame winning and losing in a compelling way (like CoLab above)?
- **Wave:** What's the name of the wave you're riding? Are there other megatrends you can tap, perhaps from adjacent disciplines?
- **Niche:** What's the name of the segment or vertical you'll position around? What about the customer-innovator behavior you've seen or the attribute you want to own? (See the **find it** chapter for more.)
- **Category:** Do you have a unique category definition (like "continuous testing" within "automated testing")? Are you trying to pioneer a particular sub-category? (Categories are still useful framing devices!) Do buyers share and understand that?

Specifics

- **Solution:** How do you describe your solution? Is it by metaphor or analogy (X for Y)? How would you ELI5 it (Explain it Like I'm 5)? What's your name for the success? How do you describe it?
- **Your playbook:** Do you have a named playbook? (Like Qualtrics and "experience management"?)
- **Product loop:** Give your features some structure: Group ideas, name sections, split things into three, and find a coherent flow. Is there a name for this workflow or product loop you've built?
- **Behavior:** What do you call it when people do more (or less) of your desired behavior when they buy into your approach? What's the old way/new way?
- **Jobs to be Done:** Can you name the top job(s)-to-be-done your customers use your product for, even if it's just by verb?

(Manage, automate, collaborate, protect, measure, verify, optimize, etc.)

Whole

- **Your key value props:** How do you describe the *value* of your core product pillars in a clear, crisp way? What specific value do people get? What's the 10x outcome?
- **User/team/company:** What are the outcomes on an individual, management, and strategic level for your customer?
- **"The promised land":** What does the buyer's career or the company's future look like on the other side of their challenge in the 'promised land'? (And have you actually seen this?)
- **Maturity:** Can you describe a maturity path you take clients through that starts with baby steps and scales to solve their biggest challenges?

That's me casting a big net for ideas, and you can see the overlap with the questions from the previous chapter. The point isn't to rehash your answers here; it's to name concepts and flesh out your narrative world. What are the core concepts that emerge?

In some ways, your job here is to be chief taxonomist — the one who names and classifies things. Ultimately that's an old way vs. a new way, but it can be all kinds of things.

I try to walk the talk, too. I've tried to name *super positioning, the hook test, radars and laser beams, story 1 & story 2*, the four strategies (*ride it, find it, own it,* and *combine it*), and so on.

The point is to build a rich narrative universe that your buyers can see themselves in. Again, you have to be careful not to try and mash them all together into some unwieldy *value blob* (another named concept!) that sinks under its own weight — we want to bring things back to a hook — but that's a lot easier to do when you've mapped the conceptual universe you and your buyers inhabit.

Your turn

- How do you describe your current narrative universe? How do

you set the context? What are your unique specifics? What's the new 'whole' you land on?
- What's your current micro-pitch? If you had to get your new micro-pitch down to 4-6 words, what would it be?
- Write down a few variations, playing with some of the above (problem/solution, wave/value prop, old behavior/new behavior, new wave/new playbook, etc.).

3. Strategic narrative outlines

I've been teasing the actual deck outline, now let's get into it. I'm going to give you a short version and a longer version. First, though, let's establish some formatting rules.

Deck formatting

I suggest starting with a doc-based tool because this is a *thinking* exercise, not a production one. There's no need to design anything beyond the absolute bare minimum for layout. All custom style is distraction. This is words, words, words.

I personally find Notion a great place to do this thinking and writing because there's *just* enough formatting for structure and nothing else to fuss with when writing, but any other similar tool would work. You can use Google Docs, of course, or if you're more comfortable in PowerPoint, Google Slides, Obsidian, VS Code, or your app of choice (I use Figma for decks occasionally, too), that's fine, knock yourself out. The tool doesn't matter much so long as it stays out of your way.

Putting together 'slides' in a document editor may sound weird, but that's how I like to do it, given that's what 'slides' at this stage are: just text broken up over a doc, perhaps with 'speaker notes' in callouts or comments. This makes it easy to scroll through, share, and otherwise take in as one end-to-end whole.

In terms of slide formatting, I suggest starting with something as simple as:

Slide header

A concise sentence or two that captures the idea and states your key value prop.

1. **Bullet one:** You can also use two or three bullets to make your point.
2. **Bullet two:** Consider arranging them side by side if that helps.
3. **Bullet three:** You want to avoid slabs of text — keep refining until it's clear.

That's it! Add speaker notes if appropriate, but that's the barebones style we want to start with.

Short version: Narrative sketch

Now, let's get writing, starting with something quick and dirty.

Here's a five-point example narrative *sketch* for a fairly typical, insight-driven B2B narrative. (By sketch, I mean this is the easiest, roughest way to get some initial ideas down.)

This is the kind of outline I use with my clients that balances story 1 & story 2. Grab a notepad or your text editor of choice and give it a go:

1. **BLUF:** Bottom line up front (use your micro-pitch!) to introduce your story and product.
2. **Context (macro or micro):** Set up your story 1 — the broader context or your perspective on the market. What's your unique market or product insight?
3. **Tension:** What's the tension that brought the prospect here? What's the obstacle stopping the prospect from moving forward? Why don't alternatives work? And what does the *positive* opportunity for them to keep succeeding in their role look like?
4. **Solution/specifics:** Story 2 — the product reveal that addresses the tension we've set up from the story 1 change in the environment, perhaps with a brief overview of the major

product pillars.

5. **Resolution/the whole:** The impact and results the solution delivers, and what the solution means for the broader team and organization. This is the outcome — the integrated whole — they can expect and others have achieved.

Pretty straightforward, right? We've got an introduction, the story 1 context (R), a transition, the story 2 specifics (L), and an outcome and resolution (R).

Your turn

- Think about what your narrative needs to do internally. Are you trying to articulate your ideas a founder to start with? Or are you trying to find common ground with the other leaders in your company?
- If you're working out your own thinking, consider coming up with variations where you stretch your own thinking further out or further down. What's the **maximal** story 1 approach you could take?
- If you're working on alignment, and as a tip, an interesting alignment exercise might be to get your leadership team to write their own sketch individually and then bring it back to the group to compare. Or, consider what a narrative sketch might look like for a C-suite pitch vs. an inbound end-user pitch.

Long version: The 10-slide outline

Maybe the five-slide version is enough; maybe you want something a bit more substantial.

Below is my 10-slide narrative outline, and hopefully this gives you a few more ideas for fleshing out your strategic narrative. Remember, these are the ideas that power your entire startup, so don't be afraid to put aside the time and really chew on them. And if you need more (or fewer) slides to fully capture you're story, that's fine! I use 23 slides on my homepage pitch, for example. 10 isn't a magic number, it's just a guide.

If you get stuck, drop this outline into a Google doc or Notion, feed it into an LLM, refer back to your strategic specifics from the last chapter, whatever you need to get started, and then think through each slide in more detail.

Open your doc editor of choice and get writing!

Slide 1: Intro & positioning

- **Lead with your hook** — what are the choice few words that capture your concept or product?
- **Remember BLUF** — bottom line up front, and don't forget the basics. What category do you operate in? What's the value you're proposing?
- **Nail your positioning** — explain your positioning in a sentence or two. Your audience should immediately understand if it's for them or not — don't make them wait to figure it out.

This is, in a sense, the toughest slide to write because it captures the most challenging choices you have to make. Don't get stuck here — you might find that fleshing out the rest of your narrative, or even reflecting on your homepage messaging (as we'll cover in the next chapter) helps you find a catchy introduction that you can bring back to this slide.

Slide 2: Macro trends/pain

- **Set up** — now we can step back and set up our narrative. Consider slides titles like: Why {product}, Megatrends, Trends, Why now, What we see, {The name of your category concept}, etc.
- **Story 1 context setting** — If you're leaning on your change story, introduce the wave, change, shift, or new reality you've observed.
- **Story 2 context setting** — If you're leaning on your tight positioning or product, highlight the specific pains you've pinpointed.

This slide should hit on an 'inciting incident' (to use storytelling jargon) that will drive the prospect's behavior change. Either that's a new understanding of the world (story 1, *à la* Zuroa) that necessitates action from the prospect, or a new way to manipulate the world (story 2, *à la* Snowflake) that solves some existing pain point. Either way, what you're putting on their radar sets the context for the behavior change you're suggesting.

Slide 3: The need/insight

- **Build a bridge** — We've set some broader context, now we want to build a bridge to the prospect's reality. Our insight only lands if it's relatable!
- **Locate the prospect** — What tension does our insight create for the prospect in their day to day? Does it require a strategic shift on their part? Does it speak to a pain or need they feel acutely?
- **Put the broader tension on their radar** — Later we'll look at what our solution means for them, their team, and their company; here you can consider extrapolating what the pain means for them, their team, and their company or broader progress. (This is what MJ Smith did with CoLab's Company A/Company B — they extrapolated from a specific problem to entire trajectories for the prospect's company.)
- **"I hadn't thought about it like that"** — At a minimum you want them to feel like you've read their mind ("That's *exactly* what we're going through!"), and bonus points if you can get them thinking about their situation in a new way they hadn't considered.

In an actual sales situation you'd create plenty of space for prospects to reflect on the ideas you're putting to them. Perhaps they'll agree, perhaps they'll bring their own nuance, either way that's ripe for further discovery and probing. The magic words are "Tell me more about that" and "Why is that a problem for you?". Buyers want to feel heard, so while we're focusing on the stuff *you* say here, listening carefully is just as important!

Slide 4: The challenge/reality

- **Name the challenge** — We've established the problem is there, and it's there for them, but *why* is it there for them? What's the reality that makes this such a common issue? Consider naming the challenge if you can (if it doesn't sound too corny). Show you've thought hard about it. "We call this challenge the _____." or "We often find that the reason that drives this is _____."

- **The decisive moment** — In a way, this is the decisive point in the narrative — either they're open to the reality you're describing and they buy in, or they don't. The strength of this slide depends on your *customer* vision — how well you've understood exactly where they struggle.
- **Three pain examples** — In practical terms, can you provide three reasons that flesh out why this challenge is so difficult, frequent, or otherwise painful for the prospect? We're trying to build some tension before the full reveal of our solution — the more we can get them invested in and relating to the problem, the more of a relief or revelation our solution will be.

It's important to note that we're not trying to doom the prospect here. We're trying to lay the conditions for building their confidence about our solution and what *they* can do to solve their problem in the best way possible.

There's a big emotional component to selling, and there's a big difference between hammering a prospect with scary YOU'RE DOING IT WRONG AND YOU'RE GOING TO BE A BIG LOSER vs. building confidence around a shared context you can move them out of.

The doom and gloom stuff is both a bad experience for the buyer and ineffective in helping them take confident action. B2B sales analysis of 2.5 million sales conversations suggests buyers are more scared of *messing up* (and buying the wrong thing) than *missing out* on your solution (to quote Matt Dixon and Ted McKenna in *The JOLT Effect,* the book that presents the research), and building buyer confidence not in your solution per se, but *in the buyer themselves* is crucial for the sale. Doomed people don't decide (they procrastinate); confident people do.

Slide 5: The status quo

- **Range of solutions** — For any given change or pain the prospect faces there are a range of possible solutions. They can do nothing (stick with their status quo), use generic solutions (throw people at the problem, use spreadsheets, etc.), or choose a peer competitor. You need to...

- **Goldilocks it** — Your job is to show why the status quo and/or alternate solutions don't really solve the core problem. I love using a Goldilocks slide here if I can, first framing alternate solutions as too hot, too cold, or too whatever (simplistic, expensive, time-consuming, cumbersome, etc.).
- **"They're this, we're that"** — Consider Ryan Yackel's approach of bucketing the competition in a way that makes you stand alone as the unique solution. This is often a handy way of dealing with peer competitors without naming them. If you're facing off against an incumbent, however, feel free to target them head on — your prospect absolutely will look at them (or have used them in the past).

Your explanation for why alternatives fail has to be believable. The classic B2B approach might be to highlight that spreadsheets are too brittle, manual processes aren't good enough, generic solutions weren't built for your specific needs/ the big change that's happening, and so on. Too hot, too cold, in preparation for what's *just right*.

Slide 6: The opportunity/how they succeed

- **Tease success** — This is a strategic narrative, so we're trying to keep things high level, and that includes success beyond merely buying the product. That means we want to put what *success* can look like on the buyer's radar too, not just their pains or change they face, which any vendor can hit if they want to.
- **"Wouldn't it be great if..."** — We want to make it clear we understand what their success is the outcome they get. Keep the framing around *their* success, their job-to-be-done (not your product!), and the relief of their pain — wouldn't it be great if you could do A and B without worrying about C? Wouldn't it be great if you never had to deal with D or E ever again?
- **The promised land:** Andy Raskin's device of teasing a 'promised land' can be helpful here, too. Again, we want more

confidence-building carrots (a clear vision of their success) rather than confidence-destroying sticks (if you don't do this, you're a big doomed loser!), so try and think about the top 3 bullets you can hit that describe that success.

Proof makes slides like this land. I won't dwell on it here because your outcomes will be unique to you, but matching a 'Wouldn't it be great if...' to 'And that's what we see, with X of our customers achieving Y in Z time' can be a powerful way to deliver that proof.

Slide 7: The product

- **The big reveal** — Now we finally pull back the curtains and reveal our solution. We've hopefully established common ground and agreement with the prospect about what their pain and potential success looks like. If they've bought into our perspective and insight, they should now be invested in what our solution might mean for them.
- **Answer the tension** — This is where we relieve the tension we've been building in the narrative thus far — we've established we know their problems, we know their real challenges, we know their alternatives don't work, we know 'it would be great if...', and now **drum roll** we've got the solution just for them. We can now get into the story 2, narrow-focus specifics.

This is the middle L in R/L/R, and in most typical sales calls, this is where you'd spend a considerable amount of time. Perhaps, like in the Arrows example above, this is where you'd spend all your time! But that's the difference between a strategic narrative and a product demo pitch — the points we've hit in the slides leading up to this part might be points you hit in social media, in founder keynotes, in content marketing, or in other outreach. Or they might capture the positioning choices you've made internally so the buyers you sell to don't need much context setting because you've tailored everything so precisely to their needs. (We'll discuss tailoring your narrative for specific sales contexts further below.)

Slide 8: The product pillars

- **Walking through the product** — Again, this is a strategic narrative not a product demo, but you might still flesh out your own narrative with the product pillars ("Dashboards," "Workflows," "AI," etc.) that make and differentiate your product.
- **From 'why' and 'what' to 'how'** — Now it's time to show *how* the product does all these great things we've been discussing. Time to walk the talk. And remember to bring features back to the value and outcomes we've put on their radar earlier!

Be sure to hit objections as you go, too. Maybe folks always ask about a Foobot feature when you know they need your Whatsit automation. Knock objections on their head, and pre-emptively if you can, as that's what high-performing sellers do.

Slide 9: The impact

- **What it all means** — What does your concept and solution mean for the team? How does it help them level up?
- **Ladder up benefits** — This slide is a great place to neatly articulate the benefits of your solution for **end users**, **the team/function** more broadly, and perhaps **the company on the whole** and the strategic challenge they were facing.
- **Micro to macro** — Consider going from micro to macro, from workflow improvements from those using the tool, to net business outcomes for management. Remember, your product isn't an end in itself; it's the outcome that matters, and complex B2B products often have different outcomes for folks up and down the org

This can be a handy place for 'from → to' descriptions, i.e., then → now, or current state → future state. E.g., front-line workers may go *from* 'struggling with X' *to* 'confident with Y,' and management might go from 'lacking insight about A' → 'full visibility due thanks to B' and so on.

Note: See the discussion below about business value vs. personal value — this is really where you want to drive home your understanding of the buyer's personal value, i.e., what will make them look and feel good.

Slide 10: The result/outcome

- **The integrated whole** — We end on the final R of the R/L/R jump, focused on a new integrated whole with your product at the center.
- **Full circle:** This is where we bring things full circle and tie the narrative back to that initial hook or three-word value prop we opened with, elaborating on the glorious future that awaits the prospect.
- **Value, value, value:** Consider using very succinct bullet points to round out this slide that describe the new successful future that awaits your prospect enabled by your solution, hitting your major value themes and introducing them with a single word (productivity, efficiency, safety, speed, etc.).

Really spell things out for folks here. You could also consider using an A → B → C sequence that recaps (a) how your product solves the challenge we raised, (b) delivers meaningful quantitative or qualitative outcomes, and thus (c) achieves the original value proposition from the very first slide.

This is both a useful way to recap the narrative *and* a blueprint for what your Customer Success team should be helping your customers achieve. That success is what will inspire those customers to tell others, driving word of mouth and building your brand as a solution everyone just 'knows' about.

Slides >10: What comes next

- **Elaborating on your narrative** — For an actual sales deck, you might include more on your product, proof (testimonials, logos, metrics, customer stories), and next steps.
- **Always suggest next steps** — Your deck will naturally end with a focus on next steps for qualified buyers — that might mean further demos for other stakeholders, answering objections, pilots, trials, contracts, etc. That's outside the scope of our narrative, but (for those new to sales) it's incredibly important to end actual sales calls with clear next steps so the prospects can either evaluate further (e.g., getting

hands-on) or otherwise make progress towards a decision. No "So... whatcha think?"

That's the outline!

That was a lot to chew on, so let me remind you that all of this is built on your hook — that simple, unique idea that captures what you're about. Hopefully this strategic narrative exercise has helped you land on a hook that's extra sharp.

A few more things to note...

Polite agreement = no

Remember, a successful narrative isn't one you've delivered; it's one the prospect has truly heard. How do you know the prospect has really heard and internalized what you've discussed? Look at what they do, not just what they say. For early-stage folks, anything less than a 'YES!' is probably a 'no', *even if they sound interested*. Early-stage founders can struggle with this. You'll get a lot of polite agreement and even half-hearted requests for getting hands-on (as sales pro Jen Abel notes) from folks who are actually unqualified to buy from you. Keep iterating and stretching your thinking until you see their eyes light up!

Hitting personal value in the sale

In 2013, CEB, now part of Gartner, did a study of B2B buyers (traditional B2B, i.e., across professional services, industrial supplies, printing, shipping, etc.) and found that personal value has 2x the impact of brand value in a sale.

First, brands need to convey business value, but that value was not differentiating *at all* (my bold):

> *Unfortunately, brands' widespread success at demonstrating business value has limitations. Although each brand can claim impressive business outcomes, so too can that brand's competitors. In fact, business value perceptions hardly vary at all between brands, either within an industry or even across industries (Fig. 5).* ***The reality is that buyers perceive little difference between the business value various suppliers can offer.***

Business value, they wrote, was "table stakes." Or, as they put it, "It gets suppliers into a buyer's consideration set but doesn't make them stand out within that set of competitors." Buyers then just pick the top, seemingly interchangeable vendors and make them compete on price.

CEB's solution was to focus on *personal value,* which they describe as:

> *[E]motional appeals in areas such as professional benefits (e.g., being a better leader, simplifying my life), social benefits (e.g., fitting in with colleagues, admiration from others), emotional benefits (e.g., confidence, excitement, happiness), and self-image benefits (e.g., doing good for society, feeling of accomplishment).*

CEB claimed that, in their data, they saw twice as much impact from a focus on personal value vs. business value. They frame this as a need to focus on "emotion" (personal benefits) vs. "logic" (rational business value). I don't buy into that framing (it's pretty logical to want better personal outcomes!), but I *do* find the results very interesting in terms of what we're putting on folks' radars.

It's obvious when you think about it — we're self-interested beings, so of course personal benefits for us as a buyer would be highly salient versus generic business value that every brand in the category offers. When we think about positioning about being as close to the customer as possible (i.e., 'there's a murderer in your neighborhood' vs. 'there's a bad person in your city'), then of course you're going to build a stronger position in the buyer's mind if you can speak to their benefits directly.

And in the AI age, it's important not to frame business value in a way that comes off as *negative* personal value for end-users or front-line staff if you expect them to adopt your platform. Pitching AI agents as a cost-cutting tool might resonate with an executive looking at the budget of their department, but if end-users think it's going to take their jobs, they're hardly likely to champion the tool's adoption. Instead, it's better to pitch the platform as something that turns them into hyper-productive badasses on a personal level while telling executives they can scale without adding headcount on a business level. That makes it a win-win-win for all involved.

Either way, how do you know what your buyers personally value? Ask them, then mirror it back to them!

Dialogues and discovery, not monologues and magic spells

Your strategic narrative is intended as a positioning and alignment tool, but remember that in a sales situation, Sales 101 would suggest several jumping-off points where you would engage with the prospect on their specific issues.

Leave some awkward silence and let the prospect agree or disagree — give them space to engage and relate what you're saying back to their own situation. (If you're early-stage, this is a great opportunity for customer development.)

Likewise, in any sales process, there will be some upfront discovery so you can weave the narrative into the prospect's situation. The strategic narrative isn't a spell you cast on unsuspecting prospects; it's a framework for discussion around their issues that leads back to your product.

Tension, tension, tension

A boring story is a series of self-serving facts with no stakes and a predictable outcome. An interesting story has tension — a relatable character you're invested in (i.e., the buyer, not the vendor!), real stakes, the plausible, real struggles they face to overcome a genuine problem, and a satisfying solution as a payoff (not fake ROI stats). That's the narrative drive that makes an outline like the above work. If you're unsure what that looks like in your context, feed your narrative to Claude and ask it to raise the stakes!

Alignment is the goal

Having a strategic narrative is one thing, building an actual position in the mind of the buyer is another.

To do that, your GTM operation needs to work together to deliver a sustained and consistent message over months and years, so don't worry too much about exact word choice or other hair-splitting, but *do* worry a lot about whether the whole team is on board with the message you're trying to deliver.

Think of your strategic narrative like a product that you need to develop along with your actual product. The same rules apply — prototyping, testing, and iterating before getting alignment and rolling it out more broadly. This is about the prototyping stage, but again, perfection isn't necessarily the aim, it's landing on something everyone can get behind and push out into the market.

Rolling out a finished deck

When you're satisfied with your narrative and you've tested it, tailored it, and found a pitch format that works for your product and audience, then what?

Then the real work begins, especially if you're in a large, late-stage company!

Ernesto Ongaro, Staff Product Marketing Manager at dbt Labs, provided some tips on what that looks like. Ernesto and his collaborators used the narrative outline above (sourced from my earlier book, *Positioning Playbook*) to craft dbt's new first meeting deck. Their GTM team knew the product inside-out, but they wanted to put together a higher-level introduction *and ensure sellers used it*. That meant:

- **Alignment:** Getting input from sales folks and sales leadership, and fighting hard to keep things simple, especially when everyone had suggestions for additions, but never subtractions.
- **Implementation:** Once the deck was approved, Ernesto had all sales managers record themselves using the deck for their teams, with the managers encouraging their reports to use it, too.
- **Promotion:** Ernesto's team captured and recirculated a lot of positive feedback from Sales that came from the new deck to encourage further adoption.
- **Competition:** Finally, they had a pitch competition where a rep from each team with the best introductory pitch was selected to win a prize.

Ernesto was kind enough to keep me updated on his progress,

and as he says, getting the deck out there took "a lot of time — building it, defending it, and rolling it out." But the positive reaction he received from his sales colleagues — and the real-world usage he could see in Gong — meant all the work paid off. That's the benefit of genuine, end-to-end collaboration and follow-through.

4. Tailoring your narrative

The narrative outline we've looked at is just a starting point, not The Definitive Narrative Outline For Any Situation™, so as an optional next step, consider how you might tailor your narrative to fit specific sales jobs-to-be-done.

We're working our way from a high-level narrative to a more practical, sales-focused pitch here. If you're satisfied with your narrative, great! You can skip the rest of this chapter and head on to the next. If you want to turn your strategic narrative into a more effective sales tool, however, keep reading.

As mentioned earlier, the strategic narrative is usually pretty heavy on the story 1 side of things. We're trying to think through all the great things we can put on folks' radars. That's fine in enterprise sales when you're pitching execs (who are using right-brain radar attention), but it can flop when pitching end-users. Different contexts mean different modes of attention, and they require different narrative approaches.

Here are a few examples.

The SMB pitch

SMB pitches might be a simple "We're X for Y" where you then get straight into the product. A little bit of story 1 — just a couple of slides, perhaps making it clear you really are for SMBs — before a *lot* of story 2 focus on how your features are super useful for your buyer, who will probably be the user. You can still put a killer idea on their radar, but the rest of your pitch is going to look very different from your strategic narrative.

SMB pitches usually mean something closer to a product-led GTM motion, and in this case, the R/L/R jump may be more of a case of:

- **R:** Putting the product on prospects' radars through mass marketing.
- **L:** Getting prospects into the product itself, or showing a very hands-on demos, perhaps as a group webinar or as a recording.
- **R:** Saving the more strategic outcomes, like what the product or platform might mean for the whole department or company if it was adopted more broadly, until after users have become customers and have already gotten value out of the tool.

The same R/L/R pattern is still there, but the PLG vs. enterprise ends of the spectrum will look very different in practice.

The executive pitch

Executive pitches, on other hand, are much more narrative driven, given it's unlikely these folks are getting hands-on with the product itself. Therefore, these pitches are more about what's *relevant*, and are built around trends and change. This can mean they look much more like your strategic narrative.

How does this work, though? Why would an executive take a sales call, anyway? Don't they have people for that? Sometimes! However, if you, as a seller, can share intel with them as someone who regularly hears from multiple other CxOs about what they're struggling with, then you can operate from a position of trust. When pitching, then, you might take them deep into the darkness before offering a ray of light.

For example, one of the authorities on modern B2B selling is Matt Dixon, who I mentioned above. Dixon is one of the original authors of sales classic *The Challenger Sale* and has gone on to write other excellent B2B sales books like *The JOLT Effect*. Dixon is not just an expert in B2B selling, he's an expert in *selling sales insights to B2B sellers*, a rare combination of both theory and practice.

In an interview for HubSpot, Dixon shared his framework for selling to CEOs and revenue department leaders — folks who lean pretty heavily on their right-brain radar. Dixon calls this approach "The Lift," but I'm going to dub it the "ray of sunshine" pitch. You'll see why in a moment.

The ray of sunshine

What would your strategic narrative look like as a C-suite tailored pitch?

Here's my paraphrase of Dixon's 8-slide framework as the ray of sunshine:

1. **The secular change(s) happening in the world:** Dixon starts with these 'mega-trends' and might rattle off several.
2. **The impact for people like you:** I.e., these trends are a wave, and it's coming for, to quote Dixon, "business leaders like you," with some specific implications.
3. **The pain it'll cause:** This is where the painful second-order effects happen, and here we dial up the pain.
4. **Stuck in the maze:** Today, "most companies" are trying to find a way out.
5. **The dead ends:** But, as Dixon says, these folks are getting mediocre results, i.e., hitting dead ends.
6. **Glimmer of hope:** However, some folks are doing things differently.
7. **Ray of sunshine:** And getting far better results than the stuck-in-the-maze folks.
8. **Blinding light:** Here's how it works, segueing into your solution.

You can think of this as a three-act narrative. In act one, darkness descends. In act two, everyone is lost and stuck. In act three, a ray of sunshine appears and turns to a blinding light as hope returns.

Now, to be clear, I made up those titles to describe Dixon's slides, but it helps highlight the flow of the story for a right-brain story 1 pitch, and that deck outline is an excellent example of what a strong, right-brain, radar-attention pitch looks like. Big trends, big implications, big stakes.

The positioning headshot

For a mid-market pitch, I'd go for something I call "the positioning headshot". This sits between the two extremes of C-suite story 1 insight selling and SMB story 2 show-me-the-solution-now selling, with a particular focus on tight positioning, i.e., the narrow-focus story 2 approach.

This format will, I imagine, suit most B2B SaaS & AI folks serving the mid-market. It's a real-world spin on the initial outline we looked at above.

This outline is based on a client I worked with several years ago, with a slight twist.

Imagine you're the seller here and your startup is a B2B SaaS with a search marketing solution for retailers with large catalogs.

Your prospects — retailers — know they need more organic traffic, they know SEO tools exist, but they don't know about yours. Therefore, we've got to both get on their radar (story 1) *and* then pivot to why this very specific approach is a perfect fit for them (story 2).

What would a pitch here look like?

1. **Name the problem:** To get on their right-brain radar, we'll lead with an insight into a problem and opportunity *for this specific segment* they haven't paid much attention to. We might give it a name like 'paid spend dependency' as a framing device and then discuss our diagnosis of missing search traffic — search traffic that *should* find them but isn't. (Note that this *isn't* a broad observation about change in the world in general; it's a relatable insight into their world right now, and the more specific, the better.)
2. **The alternatives:** Retailers could solve for this missing organic traffic in a variety of ways, but they're all bad options for a variety of reasons, and we might discuss why this hasn't been solved and why the pain persists. (Specifically, the pitch might be: Retailers can have massive catalogs; most SEO tools are targeted at manually created written content; retailers can't create relevant pages manually at the scale; it's therefore an unsolved problem.)
3. **Agitate the pain:** The traffic you should be getting is missing and significant money is being left on the table. (Ideally, the prospect can relate to the difficulties they've faced if they've tried to address this in the past.)
4. **The tension:** (This is the part most folks skip!) We might

sprinkle these initial insights with slightly awkward or uncomfortable questions about the customer's story to establish consensus on the problem, perhaps digging into whether they *really* need this solved and where they've struggled in the past. (Too many folks just accept buyer skepticism and plow on with the pitch without switching gears to take a genuine interest in the customer's actual situation.)

5. **The target:** (Note how similar the first four slides are to the Ray of Sunshine pitch. Now we pivot to story 2.) We want to zero in on this very specific problem. We're focused specifically on helping retailers with large catalogs who are missing out on search traffic, so what does this look like specifically for this prospect? We need the prospect to give us the target to aim for here, establishing consensus on the exact problem we've diagnosed, paving the way for...
6. **The headshot:** Now we go for the headshot at the target we've established, introducing our fits-like-a-glove solution for the prospect's very specific challenge, resolving all that awkward tension from before.
7. **Proof and pillars:** You then might work through product pillars to demonstrate the 'how' of your solution and the outcomes you've achieved for others, knocking down objections as you go.
8. **Next steps:** The pilot, trial, live demo, whatever that might be — just keep it as low stakes as you can to minimize buyer fear, and take them step by step on the journey towards the deal.

The right-brain executive pitch was more about selling an approach or transformation that might necessitate a tool; the left-brain SMB approach was, after perhaps a couple of context-setting slides, almost pure product explanation with an offer to close.

In this case, however, we're using our insight into the very specific challenges of a very specific group of prospects to get on folks' right-brain radar, and then we lock onto their problem with a very tight positioning-driven story 2 approach to the product. That's the proverbial headshot.

As an additional exercise, go back over your narrative and experiment with tailoring it more to an audience that makes sense for you.

5. Creating value in your pitch

If you're a bit of a sales nerd, let me leave you with a couple more ideas. This can also be a final exercise to think about how your narrative (and resulting sales pitch) can be value-creating in and of itself.

Some folks use the old 'show up and throw up' approach to selling. I guess if your product really is that new or your prospect really is that unaware, then sure, you're going to have to do a lot of explaining.

But selling can offer so much more value than that. Your pitch alone can create value for your prospect in the right circumstances. Think about who they'd rather buy from — someone who just talked at them for 45 minutes straight or someone who understood their problems, slowed down to dig in more when required, and gave them valuable insights from the get-go?

There are a couple of right and left-brain ways we can do this. McGilchrist suggests that our right brain tends to be good at pattern matching, while our left brain tends to be good at mapping things — like the market you operate in, so let's run with that.

Customer expertise

On the right-brain pattern-matching side, I like the idea that *pattern matching leads to customer expertise*.

If your positioning is tight enough, you start seeing the same problems over and over. David C. Baker makes this point in his excellent book *The Business of Expertise*. If you see the same problems again and again, you can predict prospect reactions and potential outcomes again and again. Why? Because you've become an expert. You've seen the patterns play out dozens, hundreds, or thousands of times. This gives you the vision for (and insight into) what might work for a particular prospect, allowing you to start delivering value in your pitch beyond just talking about the product.

How can your startup demonstrate that kind of customer expertise?

If you can nail this, your prospect will ideally be more inclined to trust you to solve their problem over the competition. Your positioning has given you the insight, expertise, and vision your competitors lack.

This is the "repeatable case study" B2B folks often talk about.

Speaking of competitors, you'll note that we didn't spend much time on the competition (beyond the status quo) in our sales narrative above.

Ideally, you're pitching something new and innovative and therefore don't need to address the competition beyond framing them as the traditional, legacy, old-school, or early approach that your innovation obviates.

However, the reality of most B2B tech categories is that there's usually one or more entrenched incumbents and dozens or hundreds of other challenger companies, and they get a say, too.

This can make things very confusing for prospects and results in a lot of 'no decision' deal outcomes simply because buyers couldn't reach a high-confidence decision they were happy with (as Dixon and his co-author Ted McKenna document in *The JOLT Effect*).

How do you create value for buyers in this predicament? This is where left-brain market mapping gives you market insight.

Market mapping (a.k.a. sensemaking)

As a final exercise, consider how you might bucket alternatives and give buyers a rough map of the whole market. If you can, you can resolve a great deal of their confusion (as April Dunford often talks about).

This can be especially helpful for early adopters who have been told to go out and buy one of several AI solutions that all sound the same and that they know very little about.

Here, I'm not talking about the corny 2x2 slide where you happen to occupy the sole upper-right 'good' quadrant and everyone else is in the bad/lame/expensive quadrants. Keep it believable and helpful — buyers know you're already putting your spin on things, you don't have to beat them over the head with it.

That said, if it *is* believable that you exist in a new box and the competition exists in an old box, then run with it!

Progress vs. choice

Again, notice the difference in hemispheric approaches. There are right-brain approaches that are more about putting something new on folks' radars, perhaps born of the expertise you've developed seeing the same problems they've experienced over and over again, and framing your difference in a then-and-now sense. That's the change-over-time approach. Here, you're helping folks make **progress.**

Then there's the left-brain approach of focusing down on narrow specifics and differences and analytically mapping out alternatives so you and your prospects know exactly where you fit relative to the competition. Here's you're helping folks make a **choice.**

Both approaches can create value for your prospects, so as you work on your narrative, consider to what extent you can do both.

Your turn

- Do you have specific customer expertise you can put on your prospects' radar that speaks to where they're at as they try to make progress towards their goal?
- Can you map the market in a meaningful way that speaks to the specific differences between solutions in your category?

Narrative-driven super positioning

The art of super positioning is using your collective vision as a team — being able to look out *and* down — to building a super position in your market.

It can take a whole team to get there, too, with insights from product, engineering, marketing (product marketing especially), and sales all contributing. That's why exercises that tease out these insights and drive alignment around a strategic narrative can be so helpful.

But at the end of the day, someone has to pull the insights together and turn them into a coherent vision, positioning, and strategic narrative that drives the company forward.

If that's you, hopefully the narrative tips and outlines help. If you want more help grounding your big idea, skip ahead to the strategy chapters on how you can ride it, find it, own it, or combine it. Remember, this is a choose-your-own-adventure book!

Next, however, we'll look at the final positioning exercise focused on your strategic customer-facing messaging and homepage.

10

Exercise 3: Strategic messaging

We've gone from inputs (the specifics from chapter 8) to alignment on a strategic narrative in the previous chapter (**chapter 9**), and now we'll look at customer-facing **outputs** in the form of strategic messaging that will help build your super position.

This final exercises is more of a copywriting activity where you turn your bigger narrative into some very specific web and brand copy. If your narrative was more about story 1 — what's new and why they should check you out, your strategic messaging on your homepage is more about story 2 — why you're useful and why they should choose you.

In the previous two positioning exercises, if you were working with your team, there would inevitably be a strong focus on collaboration and consensus — collaboration to tease out your strategic specifics as each function saw them, and consensus around how that came together as a strategic narrative.

This chapter, however, is more about how you creatively and concisely capture what has been discussed so far. While you inevitably need to get agreement on the homepage direction you choose, don't make this a design-by-committee exercise.

This is where you need to give your creative folks (especially your copywriter) the freedom to execute in a coherent way that — yet again — nails your specifics.

That's not to say that this is the "creative" work that happens after the "strategic" stuff is done. Far from it — strategy is about choices, and this is where you actually have to make those choices. You have to spell out the choices you've made for all the world to see on the homepage of your website. If you hedge, hesitate, or water your positioning down, have you really made the hard choices?

(This is why nailing your narrative first is so important, by the way. You can test the narrative with prospects, understand what works and what doesn't in real time, and then write your homepage with confidence.)

It's also important to resist the temptation to make your big story 1 vision front and center. Again, that's what your strategic narrative is for. Instead, this is where the folks closest to customers need to put things in the language *of* your customers. No exceptions.

With all that in mind, to get you started, here's a simple customer-facing positioning and messaging exercise to get you (and your team) thinking.

Five-minute messaging

Let's look at your website for a quick example of how you can apply your positioning and the specifics you've landed on.

Open up your homepage and look at your hero. Now think about:

- What's your category?
- What's your ideal customer?

- What value is most important to them?
- What do they replace when they choose you?

Are those things obvious in the first scroll or two of your website? Would a prospect recognize themselves in what they see?

Now, to be clear, this is copywriting 101 — it's not really market-unlocking positioning, much less super positioning.

But this is a positioning crash course after all, and it's a handy exercise to see how you can connect some of these big questions we've been talking about with the words prospects see on a page.

Big ideas; tiny bandwidth

At the end of the day, for your customers, that's what this work amounts to:

- **10 slides:** Hearing your big idea through the equivalent of 10 slides from your strategic narrative.
- **10 scrolls:** Or scrolling through 10 *sections* (max!) on a web page.

(Or, more likely, just hearing a simple hook from their colleagues or from your marketing!)

That's a very limited amount of bandwidth, which is why it's so important to nail your story 2 specifics in your messaging, especially on your website. And if you can, even in the most basic terms, simply teasing out some of those specifics can help you come up with a homepage hero that packs a punch.

Your turn

- Does your current homepage hero section answer those questions above?
- If you strip away the design and imagery, what *does* your homepage actually say? How specific is it?
- If you do the same for your peer competitors, what specifics are they hitting? What does that say about their deeper positioning strategy?

Cluely vs. Granola

How do other startups use specifics in their homepage heroes?

Let's look at two similar buzzy AI note-taking startups, Cluely and Granola.

The stunt-marketing geniuses at Cluely (no shade, that's how Salesforce started) have, as of writing, this as their homepage hero:

- **Headline:** Invisible AI That Thinks for You
- **Subhead:** Cluely is an undetectable desktop app that gives you the answers you didn't study for in every meeting and conversation.

Meanwhile, Granola, a popular app in the same space, has:

- **Headline:** The AI notepad for people in back-to-back meetings
- **Subhead:** Granola takes your raw meeting notes and makes them awesome

Which one do you think is stronger? Cluely is leaning heavily on their story 1 big idea ("AI That Thinks for You" and 'cheating' by not studying), while Granola is more focused on their story 2, X-for-Y positioning specifics ("AI notepad for people in back-to-back meetings").

Remember, your story 1 drives people to your website, and your story 2 helps them choose you. Here, Granola's approach is more specific, and therefore gets closer to my needs if I'm their ICP.

Again, closeness wins, and I can more easily position Granola in my mind because I know what it is ("AI notepad"), who it's for ("people in back-to-back meetings"), and what it does ("takes your raw meeting notes and makes them awesome").

That's the power of specifics.

Cluely, on the other hand, isn't as clear about who it's for (meetings... but also conversations?) or what its value is (do I really want an AI that thinks for me? Is that a meaningful hook?). It also has a vague offer about "answers [I] didn't study for" — but who is studying for answers in a meeting?

(I know, I know, it fits with their cheeky "cheat on everything" shtick — stay with me!)

For me, based on this little homepage hero test, it's hard to understand what position Cluely are trying to build in the mind, simply because it's much less specific. It doesn't feel as close as Granola.

Reaching the mind

But here's the twist: Cluely are going for reach, not niche. They're trying to get close to more people, even if it's not *as* close as Granola.

This is the classic niche vs. reach tradeoff, and this is where what matters isn't the individual webpage — it's the *GTM system* you're designing.

In this sense, Cluely gets points for being *very* specific about their brand-building system and the very considerable reach it generates. That is, they're being specific with their *strategy*, even at the cost of the specificity of their *messaging*.

They (in)famously use controversy ("cheat on everything!") and stunts, including a Times Square billboard that reads:

> *hi i'm roy im 21 this was very expensive pls buy my thing*
> *cluely.com*

...in stunningly vanilla centered-black-text-on-a-white-background.

No marketer who wanted to still be employed tomorrow would get away with proposing that, which is why it stands out (and why it's the founder that can pull it off).

And marketers looking in from the outside still approach it with a direct-response mentality of 'does it convert?' but that's the wrong question.

The right question is: have they built a coherent system?

Cluely, for their part, have a specific goal for their system in mind: to drive short-form videos about their product and company. It's no secret — Roy, the aforementioned founder, spells it out on X all the time. It's those second-order effects that then drive interest in the brand. The company's antics *become* the story. That puts them back on people's radar in a story 1 sense.

'We're the cheeky startup that wants to help you cheat on everything!'

This tends to drive traditional B2B marketers up the wall because, again, they're so used to the middle ⬇️ positioning prong ('It doesn't say what it is!' 'Who even is this for?' 'It must convert terribly!'). But those objections completely miss the point.

In fact, thinking *only* in those terms is probably a severe restriction on your ability to build a super position.

Positioning is more than *messaging* specifics or *first-order, direct-response* specifics. That's all the product of narrow attention. You have to consider the whole you're building, too, as we discussed last chapter when looking at the R/L/R jump.

That final "R" means brand — that integrated whole where you are the best choice for Y. To build *that* position in the mind of the buyer, you need to reach those minds, and Cluely are very good at that part at least.

Let me be super clear that I have no idea if Cluely's approach will work — people have to actually use and like the product, too. But they're very specific about the system they're using, and the attention that generates is undeniable.

(Well, they *were* very specific about the system they're using. About 5 minutes after publishing this book, Cluely pivoted their messaging to vanilla — and I mean *vanilla* — B2B. Their website hero headline became "#1 AI assistant for meetings" with the subhead "Takes perfect notes, answers questions in real-time, and makes you the most prepared person on every call." I'm sure they'll have iterated many more times by the time you're reading this, and perhaps by then they'll have their story 1 and story 2 approaches working together!)

Super positioning through superior distribution

Let's stick with distribution for a moment, because it shapes what you say across your customer-facing assets.

Building a super position often involves unlocking distribution in a new or novel way, and Cluely's ability to unlock short-form videos is one variation on a theme that has been a part of B2B go-to-market (GTM) for a long time.

For example, HubSpot used agency partnerships, Loom used the viral growth loops that came with video sharing, and Squarespace tapped into an emergent mass channel (podcast ads) to just kind of be everywhere, all the time.

We're not going to focus on the distribution side of GTM in this book — we're more interested in the fundamentals of your narrative and messaging. Getting those right, however, helps you know *who* to sell to and how you might reach them.

That is, specific positioning unlocks — and needs to be matched by — specific GTM motions. Tight positioning unlocks tight, sales-based distribution. If you're an early-stage startup with limited resources, knowing who to focus your limited sales resources on is essential. And broad, brand-driven positioning like Cluely requires broad, social-media-driven GTM. (If that's what you're good at — and they are — then you almost *have* to go broad.)

So when you think through the three big prongs of super positioning (➡️⬇️🔄), also think about how it ties into your GTM motion and distribution, too, and be careful before you dismiss a startup or competitor for being too broad — they might be playing the story 1 game.

Validation before distribution

Let's get back to the exercise and your customer-facing messaging, especially your homepage.

You've got a narrative — that's awesome. Now, how do we turn that into something that builds a position in the mind of the buyer?

First of all, make sure you test it on prospects before steaming ahead with a full-blown rollout. You won't know if it truly works until you've put it in front of folks and watched their eyes light up (thanks, Wes Kao) *and* pulled their wallet out, metaphorically if not literally.

Remember, it's behavior change that matters at the end of the day, not just a mildly interested response or polite agreement with what you're saying. We want people excited and ready to do things differently with our product and playbook because we truly are insanely different.

Once you're confident you're on the right track, it's time to expand your experiment to your major marketing touchpoints, the most visible (and contentious!) of which is usually your homepage.

The job of your homepage

We've talked about hooks and narratives, but let's reset for a moment here and think about just how blunt you have to be to capture someone who's giving you, at best, 30 seconds of their attention as they scroll your homepage (and not 30 minutes of a pre-qualified demo call).

Let's set the scene with this tweet from Bryan Casey, VP of Inbound at IBM:

> *Across dozens, maybe hundreds of distinct pieces of user research, the most consistent piece of feedback is to literally just say what the dang product is and does. Forget positioning. Benefits. Use cases. Client pain points. At least up front. Job 1: what the hell is it?*

"What the hell is it?" indeed, a question far too many SaaS & AI homepages have failed to answer in the past.

I also like to remind folks that your homepage doesn't have to do everything, either. Product and industry-specific solution pages exist, too. Sometimes, I'll tackle a set of pages where the thrust is:

1. **Homepage:** *What* is the value you offer.
2. **Product:** *How* your product delivers that value.
3. **Solution:** *Why* that value matters for a specific industry, role, or use case.

How do you currently structure your site and messaging?

Your turn

- Briefly review or audit your web messaging. How "story 2" is it? How clear are your product descriptions? How specific is

the value you describe?
- Does your current messaging follow on from the ideas you fleshed out in your narrative, or does it need a significant revamp?
- What sort of messaging do customers see across your ads and other forms of outreach that might drive them to the website? Is there a consistent thread through your organic social, paid ads, homepage, and narrative?
- What content do people who actually become customers consume on your website? (This is an attribution question, and attribution is notoriously difficult, but at least having a sense of what **buyers** — not random visitors, actual buyers — actually read/scroll can give you a sense of what sections and pages are actually seen!)

Outcomes-first messaging

Here's a quick guide to applying your narrative to your homepage.

Homepages have to be brutally concise — 10 scrolls ain't much. And the second most important question they need to answer (after "What is it?") is "What's in it for me?"

That means leading with value and outcomes supported by the nuts and bolts of how it works.

Again, like your narrative, how good that messaging is depends on the ingredients you have. Hopefully, after the narrative exercise, you've got a lot of solid messaging ingredients to work with. Now it's time to marry those with customer outcomes — quotes, numbers, and, of course, logos. That's your proof, and the more tangible, specific, and quantifiable it is, the better.

When it comes to these pages, specifics really do sell, and the best way to ensure you're hitting those specifics is with a checklist.

So here's our first checklist, one for your homepage:

Homepage messaging checklist

- Does your homepage tell readers *specifically* what your

product is and does? (Seriously! So many SaaS sites fall at this very first hurdle. And no, "Save time and make more money" doesn't count.)

- Does your homepage tell prospects why they should choose you over the competitive alternative for their job to be done? (That may be the status quo like spreadsheets, or it might be a specific incumbent or peer competitor.)
- Does your homepage match your GTM motion? (High-awareness PLG products may push users into the product ASAP, sales-led folks often need to do more explaining upfront.)
- Does your hero section capture the BLUF (Bottom Line Up Front) value proposition of your first narrative slide, perhaps in 3 short bullets?
- Does your homepage show or visually represent the product in some way? (That's what you're selling after all — don't get so caught up in benefits or abstract imagery that you neglect the product itself.)
- Does your homepage capture a single, key, compelling reason to buy? (What's the hook? Have you validated that with customers?)
- Do you connect that value prop with how the product works, perhaps as a three-step 'how it works' or some sort of flow from start to outcome?
- Can you drive the point home with from/to, before/after, or with/without language to contrast the status quo or incumbent with your new way?
- Do your testimonials and social proof reinforce the core message you're trying to communicate in a crisp, concise way?
- Do you land on a big call-to-action (CTA) 'closer' that tries to get folks to take a clear next step (usually a demo)?
- Do you have a secondary CTA for folks who aren't ready to commit to a demo at this stage? (Think: short explainer video, mini demo video, etc.)

There's a lot of goofy The Ultimate Absolutely Guaranteed High Converting SaaS Homepage Template™ fluff out there, which is premised on the fact that if you somehow take a (very biased) average of what the SaaS industry does with their homepages and make it even more generic, somehow that will be high converting.

There's something to be said for templates to ensure you're covering the bare basics, but beyond that, what "converts" is meeting *your* prospects where they're at with *your* specifics that are relevant to them.

"Can we test that?"

If you're proposing customer-facing changes, someone will inevitably suggest we "test" the changes.

Founders and marketers alike can have very different definitions of what that means. Sometimes people will talk about "testing" in a completely vibes-based way. They mean "Let's roll it out and see what happens" in a very informal way over the months ahead. And that's fine — that's often as good as you can do in startups. There's almost never enough traffic to meaningfully A/B test variations, especially in low-volume/high-value situations. (And even if you can A/B test, you probably shouldn't unless you can measure the cost and outcomes of the testing program itself.)

Nevertheless, a select few people have visibility into a broad range of A/B test outcomes, and one of those people is Casey Hill from DoWhatWorks. On LinkedIn, Hill writes (lightly reformatted):

> *Why do your website copy tests rarely move the needle? It's an important question. Studies from teams like Optimizely (as well as our internal data) show that a large portion of copy tests have minimal impact. So what are the 10% that are effective doing differently?*

Let me interject here — Hill's observation gets to the heart of what we're trying to do with your messaging here, and it's **make an impact.**

Some folks like to say your positioning is what you put on your homepage, and while there's some truth to that *if the strategic specifics underlying your business have changed.*

If they haven't, tweaking copy *alone* is not "positioning," and is unlikely to have much impact at all. Unless, as Hill writes:

> *A lot of it comes down to whether your changes check one of these two boxes:*
>
> ✓ *Provide new information*
>
> ✓ *Provide clarity*
>
> *Good headers and subheaders, for example, clearly articulate what the product does.*
>
> *Good:* "Shop securely and choose to pay in 4 interest-free payments, in full, in 30 days, or over time."
>
> *Poor:* "Stop chasing transactions. [brand] turns customers into diehard fans—obsessed with your products, devoted to your brand, fueling your growth."
>
> *Good:* "Easy-to-use corporate cards, bill payments, accounting, and a whole lot more. All in one place."
>
> *Poor:* "An all-in-one finance solution to cover everything from HR to accounting."
>
> *The poor versions leave you with more questions than answers, while the good versions clearly state specific capabilities.*

Those examples are very helpful — the "what's in it for me" in the good examples are both strong and clear.

But I quote Hill here because this comes back to the core theme of this book: super positioning is about putting something new on folks' radar and/or giving them a clear reason to choose you by being specific about what you do and how you're different.

That's the two modes of attention again, and I think it's kind of amazing that this principle scales from the very micro — headline tests on a webpage — all the way up through how positioning is done (go big/go narrow) and the fundamental ideas that drive your startup.

Your turn

- Go back over Hill's examples and note the difference between the good and bad examples.
- How does your current messaging compare? Would you be a 'good' or 'bad' example?
- How can you move from vague language to more specific value props?

Anatomy of a professionally written B2B homepage

With that in mind, let me give you a sense of how I approach homepages as a copywriter-turned-positioning consultant.

Here's an amalgamated summary of three recent projects, to give you a taste of what works in the real world. This is the hands-on part of this chapter — see if you can take your specifics and narrative and flesh out your homepage following this structure (or be lazy and ask Claude to do it):

Hero

- An eyebrow (i.e., a tiny heading label) for the category or some context setting.
- Headline — the core hook we've landed on.
- Subhead — a brief product introduction that hits either three key value props or your chosen vertical(s).
- Bullets — the top three value props for your offer (if you didn't mention them in the subhead).

Pains block (optional)

- 3x bullet blocks (an icon, small header, and two lines of copy) about the major pains your prospects face.

How it works block

- A 1-2-3 step set of bullet blocks that describes the core product flow.

Product exploration

- A set of pillar features, or
- A set of use cases, or
- A set of broader product areas as tabs or gateway blocks to specific feature pages, depending on the complexity of your platform.
- Either here (or below) there might be gateway blocks to explore industry-specific solution pages, too.

Social proof

- A hero testimonial that echoes your value prop, lightly edited for clarity and conciseness, with the relevant phrases bolded for emphasis.
- Supporting metrics or outcomes.

Outcomes

- What it means for the users, team, and/or their customers.
- Further social proof that people have had with your product or playbook.

Integrations/where it fits

- How the product works with what you've got *or* replaces other tools.

'Whole product' support

- It's not just about the literal software, it's about the services, ecosystem, partners, and other aspects of your business that come together to help your customer succeed.

Closer

- A final call to action that calls back to the hero value proposition but encourages readers to click the big "Book a demo" or "Get started" button.

That's it! That's your 10-or-so scrolls worth of information that hopefully helps as many people as possible take the next action (sign up, book a demo, download the app).

PLG vs. enterprise & story 2 vs. story 1

Remember, as we discussed earlier, you'll probably lean less on story 1 narrative and more on story 2 product specifics here.

Products where the user is the buyer are generally going to lean very heavily on the product experience itself — maybe so much so that you're encouraged to dive right in, and the product onboarding does more of the messaging heavy lifting.

More enterprise-style offers, on the other hand, are going to have story 2 specifics incorporating references (i.e., social proof and case studies), the whole product of support, resources, and the broader ecosystem, and the big picture outcomes executives *and* end-users are looking for. That is, the pitch is built on reassurance as much (or more than) it's built on innovation.

Or, to borrow a quote from former Microsoft executive Steven Sinofsky:

> *Enterprise software is insurance. Productivity software is personal productivity.*

Pitch accordingly.

Feed the machine — prompt you, prompt me

Can your LLM of choice do this? In form, sure, if you can prompt it appropriately.

I've found chatbots actually pretty handy at creating and running through frameworks and checklists, so feel free to copy & paste the checklists and tips here into Claude (or whichever current bot has non-awful prose) and see what it spits out.

I call this the "prompt you, prompt me" approach. You prompt the AI with a framework, it creates something generic but okay-ish, that prompts you to redo it so it's actually good, and you're done.

I like working with LLMs in this way because they're good at handling the known stuff, while your whole job is to handle the unknown stuff — the novel, out-of-distribution insight that drives your company, product, and positioning. LLMs are just lossy encyclopedias, after all. They're a fantastically useful normal technology, but they don't know what's unique about your customers or what's unique about you.

That is, again, your job. They ain't automating that. You, in that sense, are the LLM that matters, just a gooey biological one, and your brain contains the model that matters — the one that can see things that are well outside the distribution of training data the LLM was built on.

Even still, it can be hard to go from that model in your mind to concise web copy. *You* need prompting. And what better to prompt your specifics than the generic-in-substance-but-accurate-in-style responses you get from an actual LLM!

The value is in the vision

All of that is to say that being able to write concisely in marketing speak isn't really the bottleneck here. It's a skill, don't get me wrong — I spent years mastering it for a reason. But ultimately, what matters is conveying those core positioning concepts you developed in your narrative, your hook, and your vision more broadly. *That's* where the value is, and if AI helps you get there, great.

Just don't confuse form and function — AI can't do your thinking or 'seeing' for you, so calibrate your expectations accordingly.

Branding & your messaging

Let's wrap up this run of positioning exercises with one last idea:

Message. %$^&ing. Discipline.

If positioning is about building a position in the mind of the buyer, then you *must* maintain absolute clarity and consistency in your messaging.

It's very easy for messaging to go off-script in the name of "testing" or having sales or marketing folks just making it up as they go, because they don't appreciate the fundamental position the company is trying to build.

By all means, test your pitch with prospects and test your ads within your various channels, but when you've settled on a message, put some guardrails around it and **lock in**.

And if the team pushes back, find out why. Maybe they don't fully believe in the message you're trying to push? If so, don't ignore them. Your messaging has to pass people's BS filters, and if folks internally don't buy it, why would prospects?

Your brand messaging doc

Ensuring messaging consistency isn't exactly the sexiest side of being a marketing leader or founder or CEO, but what's the point of all this vision and positioning work if prospects fail to hear a consistent message?

Therefore, at the end of a series of exercises like this, I like to create a simple brand messaging doc where I essentially cut & paste the messages we've landed on with some crucial context that describes *why* we're going with these particular messages.

This document should be both a reference and cheatsheet for how you talk about your:

- Category
- Product (as a headline).
- Product (as a sentence or two).
- How it works (one sentence).
- Your value proposition (perhaps 3x bullets).
- Your key messaging themes (these might be specific outcomes, where you fit in your buyer's world of tech stack, or your specific playbook).
- Industry-specific language (if you sell into multiple verticals).
- Old ways/new ways of articulating your value (especially if you're making a big shift in positioning).

Trying to do this from scratch at the start of a project can be tough, but with your narrative fleshed out and agreed upon, and perhaps your homepage copy and wireframe signed off, this should be a pretty straightforward, downhill ride.

I like to keep this pretty simple, but the complexity of your messaging docs will obviously depend on whether you're working with a 10-person startup or a 1000-person enterprise.

Some folks like building elaborate 'messaging houses', but I like to remember **the only messaging that matters is what buyers see.**

Therefore, I always try to derive it from live assets as much as possible. Messaging houses and docs should only be considered snapshots of where you landed in a point in time. Hoping a document will manage a team to be consistent is (unfortunately) a pipe dream.

That's still your job, and I hope this chapter has given you some inspiration to nail your messaging and start building that (super) position in the mind of your buyers and market!

Next steps

That wraps up our hands-on super positioning crash course. We've covered a lot of ground, including:

- Your strategic specifics as positioning inputs.
- Your strategic narrative as positioning alignment
- Your strategic messaging and homepage as positioning output.

High five if you completed any or all three steps. I'd absolutely love to see where you landed — let me know: **luke@superpositioning.co**.

If you're enjoying the hands-on approach and want to keep thinking through the strategic specifics that drive your positioning, there are plenty of ideas and examples to come, so keep reading.

Next, we'll kick off our strategy section with a look at the super positioning strategy for being super relevant — riding a wave.

PART IV

STRATEGIES & EXAMPLES

11

Ride it & prove value on a wave

Welcome to the strategy section! This is where we'll run through the four super positioning strategies — ride it, find it, own it, or combine it.

I'm going to keep throwing different positioning angles at you over the next few chapters along with more questions to get you thinking. Don't get overwhelmed — the point isn't that you need to know every angle or use every strategy, the point is that you need to find the one or two that can be a strategic specific for you that then power your narrative and messaging as you keep building out your super position.

In this chapter, we're going to start with riding a wave and proving new value, and I'll touch on Jobs to be Done and the niche vs reach dilemma, too.

I'll also provide a little summary of the strategy at the start of each of these chapters, as well. Here's the first:

Ride it in a nutshell:

- **Goal:** Find a way to be **super relevant** to your audience.
- **Strategy:** Prove your **new** value on a new wave, be that a new technology (AI), an emerging role (a new corporate function), a cultural movement ('creators') or an external development (e.g., new regulation).
- **Attention:** Active right-brain (looking out).
- **Action:** Prove it — lots of folks try *catching* a wave; the ones that succeed prove they offer new value.
- **Position:** You become the famous solution associated with the wave, often as much for your playbook as your product.

As I write, it's hard to imagine there's any other wave worth talking about than AI, so we're going to talk about AI. And nothing shows the power of a wave creating insane relevance quite like AI either. I mean, how many barely working products were adopted and given a fair chance simply because the were LLM-powered?

While AI everything might be all the rage, AI is still a tech wave like all the others that have come before, and those previous waves — including B2B SaaS wave-riding success stories — have plenty we can learn from, too.

Waves drive success

As a founder or marketer, we want to believe the success of our product in the market is down to our genius and effort.

It's a bit jarring, then, to reflect on just how fundamental *wave riding* is to the success of many great startups and B2B businesses.

Sometimes, timing really is everything. Waves of change are inherent to technology itself — each new wave of innovation drives a new wave of application, and in B2B SaaS, those applications can be worth anywhere up to, in Salesforce's case, hundreds of billions of dollars.

To understand how that works, though, let's first understand how technology waves and automation work and what the point of B2B tech fundamentally is.

Waves, automation, & productivity

Technology is a bit like sex in that every generation thinks they're the first to invent it. With the AI wave, for example, our generation seems to think this is the first time a major technology might replace a lot of the work people do, but that's been the case for every major generation of technology ever.

In 1900, for example, 40% of the U.S. population **lived on farms**. Now it's closer to 1%. Why? Science and technology have allowed us to produce far more with far less labor. That trend of abundance from less work led influential British economist John Maynard Keynes to speculate in 1930 that we'd soon be a leisure society working 15 hours per week.

There's been no shortage of AI leaders saying much the same thing — *maybe this time will be different*. Maybe this round of technology and innovation will finally Solve Work™ like mechanization, or electrification, or computers, or the internet before it. Maybe we can then *finally* embrace a capitalist leisure society, or fully automated luxury communism, or whatever else lies in wait in this technology-powered utopia.

Or maybe AI is, per Arvind Narayanan and Sayash Kapoor, just normal technology.

In any case, utopian dreams spring eternal, but I suspect we'll be waiting a while for utopia to arrive.

But consider this from another, less messianic angle.

Maybe we are already well on the way to *productivity* utopia. After all, most of us don't live on farms anymore. Technology *does* automate work, even if it doesn't *eliminate* work. We're extremely lucky to be able to work in relative comfort, on or off a farm, thanks to that fact.

That automation process will march on. There has been and will be substantial amounts of employment turbulence as managers test the waters to see what can (and can't) be done with AI and whatever comes after. Maybe that will hit harder and faster than previous technology revolutions; maybe it'll look just like them.

But life will go on, and to get there, a heck of a lot of B2B products will need to be built. Maybe you're the one to build them. Maybe you already have. Maybe this AI revolution only takes us so far, and it's back to AI winter before the next boom. But having models trained on almost all of humanity's recorded knowledge available to anyone as an API call is an insane product opportunity, hallucinations and all.

Either way, technology automates. It gives us leverage. It helps us do more with less. Maybe one day we'll finally determine we have enough and spend all our days making music or art or scrolling TikTok or whatever, but in the meantime, there's a whole lot of B2B building to do, and hopefully, this chapter gives you some clues on what that might look like for you.

Energy waves

With each new wave, it's up to us to prove — and I really mean *prove* — new value in a real way.

In theory, positioning on a wave is great because the wave does much of the heavy lifting for you. It's the groundswell of energy that you can tap that can take you far.

And a wave is a great metaphor: go too early and there's too little momentum there; wait too long and it will pass you by. Or, pretend it's not there, and risk getting smashed by it. Waves are powerful things, after all.

Waves in B2B include:

- **New general-purpose technologies:** AI & LLMs, what blockchain hoped to be (but wasn't), mobile before that, the internet (which enabled SaaS itself), and so on.
- **New corporate disciplines and roles:** Think about the rise of customer success (Gainsight), digital product design (Figma, Linear), digital marketing (HubSpot originally), or even 'creators' on the micro-business end (ConvertKit, Patreon). New roles require new tools.
- **New playbooks and trends:** There might be a move towards

agile (Jira, originally), no-code tools (Webflow, Airtable, Notion), messaging (Intercom, Drift), the shift to remote work (Zoom, Loom), or 'revenue ops' to manage GTM complexity (Clari, Gong, et al).

Your turn

- We covered some of these when thinking through your strategic specifics. Are you also riding one of those waves?
- Are you getting the most out of it, hitting the new disciplines and playbooks that emerge, too? Or is there room to maximize the energy you're getting from the wave?

Concept-wave

Those successful startups didn't just ride a wave, they often sold new *concepts* enabled by the wave. I've got a little formula for this approach, and it's:

New concept = new change + new playbook + new tool.

You might think of this as creating a category, but I find a lot of the talk about category creation involves folks getting a little too high on their own supply. I find thinking about change + playbook + tool much more concrete and prescriptive instead.

To sell your new concept, then, you need to:

- **Have the vision:** You pick the wave of change you want to ride based on what you see. The more unique your vision and the faster you can act on it, the less likely you'll be fighting it out with 100 other competitors who all saw the same thing.
- **Find the hook:** Ideally that's a change that's happening in the world — one that's *separate from you and your company* — which can be the foundation of your right-brain narrative.
- **Build your super position:** You talk about the change and playbook perhaps more than your actual tool, while building a tool that fits this new reality like a glove. This gives you a strong right-brain story 1, a prescriptive story 2 (the playbook and product), and a brand memory association you can own.

If you can ride a wave and build a super position in the market in this way, awesome. If you can't, though, don't fake it.

Don't fake the wave

One of the worst strategic choices I've seen founders and marketing leaders make is trying to fake the wave, especially in the category creation sense.

The point of the wave is that the bottom-up energy carries you to fame and fortune. If there's no energy there, though, what are you doing? What's going to carry you?

This is classic cargo-cult behavior. It's a bit like believing that if you splash around enough in a placid lake, a wave will suddenly materialize underneath you and everyone will be very impressed. I think most folks would just look at you like you're slightly deranged.

This might sound a pedantic point in the 2020s, but in the late 2010s there was a real craze around coming up with made-up new categories of software. Perhaps it was just a symptom of the tail-end of the classic B2B SaaS ZIRP era, but many VCs *really* believed it — they bought into this approach entirely. Capital was abundant after all, so why not try? The problem was that fundamental shifts in technology were less so (crypto being a B2B dead end), and founders were left trying to wish new categories into existence without a wave to drive them.

Don't get me wrong — new categories *do* happen, and maybe you'll pioneer one. That's great, and there's nothing wrong with having a fresh concept (as we'll see below) or picking an attribute (as we'll see in the next chapter) and trying to be *more about* something than the competition.

But fake category creation — and fake narratives for that matter — are a killer because the ones who believe it the most are the ones who are talking about it the most: *you and your company*. This cuts you off from seeing what's real, from meaningful learning, and from genuine market feedback because you're too busy trying to splash around to make that wave in a lake happen. But if there's no wave, it just ain't gonna happen. You're in a lake.

Don't forget to prove it

The other failure mode is to believe a wave alone is enough. You still need to solve a problem. Lots of folks will try to catch the same wave — big waves aren't secrets, hence the explosion of AI startup funding — but very few will prove their value in doing so.

That may seem painfully obvious, but it didn't appear to be very obvious in the crypto cycle, for example, where almost no one proved any meaningful B2B value. (Unless your business was crime, that is.)

You have to use your full spectrum of attention — looking out at what's new *and* down at what customers actually achieve — to develop the proof you can put back on folks' right-brain radar.

But how do you actually do that? Let's take a quick detour into the world of Jobs to be Done.

Proving it with Jobs to be Done

The tech industry is great because every day feels like day zero. Wake up, check Techmeme, scan the news, scan the feed, see a whole bunch of new products, AI models, funding announced, get breakfast.

But that day zero effect can distract us from the reality that SaaS alone is decades old now, with "high tech" going back several more decades, and there's plenty we can learn from.

That includes founders who had the vision, rode the wave, and got the pot of gold at the end (and we'll get to them, I promise).

But it also includes ideas from people who spent a lot of time wondering how to actually do the whole innovation thing in a more predictable, less random-walk kind of way.

The framework they landed on — and the startup world adopted — is called Jobs to be Done, or JTBD.

JTBD has gone through several hype cycles and comes in and out of fashion, but at its core are some very simple ideas that help us go from big vision on a wave to discrete, identifiable productivity improvements that we're aiming for.

That is, it helps us go from that big right-brain change story down to the more micro left-brain story focused on specifics and tool use.

The simplest way to express that idea is contained in the name of the theory — *focus on the job people are trying to get done.*

That is, in B2B especially, very few people wake up in the

morning wanting to play around with software, much less go through a months-long buying process. They don't want to have to think about software at all. They just want to get a particular job done.

That job-to-be-done often gets described as a 'use case' or a 'customer story' and so on, and that's fine. But the magic of JTBD is thinking about the job *behind* the job so to speak, as that's fertile ground for innovation. Use cases tend to be very literal — what customers say they want. JTBD tends to be more about what they actually need.

For example, a customer might say their use case is that they "need to see certain data on a dashboard." You can take that literally and give them what they want, and maybe that's fine, but it's not much of an innovation. A JTBD approach would be to investigate what the actual job the data and dashboard exist to serve. Maybe the user needs to take some action when the dashboard says so? Maybe an LLM-driven agent could now do that for them instead? Maybe there's some broader workflow that could be automated, enhanced, or otherwise improved?

Or, to put it in simpler terms, imagine a person who wants to hang a painting on a wall. Their use case, in terms of the tools they need, might be to hammer a nail into the wall. But from a JTBD point of view, we might ask: how else could they hang a painting? How could we innovate there? What if we had detachable sticky hooks instead? That's what 3M brought to market, and renters everywhere rejoiced because they solved the core JTBD in a whole new way.

You can see the different kinds of vision and left-brain/right-brain attention at work here. It's easy to see customer needs in a very literal sense, and sometimes that's all you need. Sometimes it makes sense to build exactly what they ask for because they know best. But once folks already have a proverbial hammer and nail, what comes next? What can you see, in terms of innovation, that could be 'out there' on the horizon instead? What's your vision for the actual job they want to get done — one that's stable over time — and not just the slight improvement to how they already do things?

(If you want to nerd out more about JTBD and positioning, I explore the topic more in *Positioning Playbook*, and we'll touch on it in the next chapter, too. If you want to go further, I recommend the parents of the theory, and JTBD has several. Tony Ulwick gave us the original, more left-brain, engineering-driven, map-and-quantify approach. Bob Moesta developed a more right-brain, narrative-driven, change-over-time approach that involves deep interviews. Most famously, however, the late, great Clayton Christensen of 'disruptive innovation' fame put JTBD on the map in 2003's *The Innovator's Solution*.)

Wave riders — AI examples

Let's look at how the AI wave is playing out in one particular area: legal tech. This is a world of masses of words — documents, cases, contracts, etc. — which makes it a juicy target for people making AI tools.

AI waves & legal tech moves

Let's touch on three different legal tech startups with three different strategies to see how they're trying to ride the AI wave, going from broad reach to narrow niche.

(Note that the funding amounts, positioning, and, heck, even existence of these companies may have changed by the time you read this. But it was an ideal time to take a snapshot of an emerging wave-driven market segment!)

Hebbia

Hebbia, targeting investment banks and legal firms (amongst others), raised $130M to be the "AI platform for knowledge work." (This is a pivot after an initial search engine play.) Their homepage boasts of "1000+ live use cases, driving real business outcomes."

This — at the time, at least — was a big-time brand-building move for reach. (A thousand use cases!) They were operating across two very potentially lucrative sectors (investment banking and law) with an apparently strong desire to expand beyond those verticals and become the *horizontal* platform for knowledge work.

Harvey

Taking a strictly vertical approach, on the other hand, is Harvey, which raised $100M calling itself the 'trusted legal AI platform.' Its product page boasts of 'a suite of products for all practice areas.'

In the vertical SaaS playbook, the vertical provides the niche, but the play is to then try and satisfy as many jobs-to-be-done as possible within that niche. A bit of niche — one specific vertical, and a bit of reach — products for 'all practice areas.'

Spellbook

At the more niche end of the spectrum (and more under the umbrella of our 'find it' positioning choice) is Spellbook, which raised $30M. Their homepage lead with "Draft and review contracts 10x faster with AI" and they claimed they were the "most popular AI tool for transactional lawyers."

As we touched on in our strategic specifics chapter, Spellbook had identified a specific job-to-be-done they can do 10x faster ("Draft and review contracts") *and* picked out a target ICP this is most relevant for ("transactional lawyers").

This is the classic wedge play for a point solution to get into the market and build momentum through extremely tight focus.

This is just one small slice of the world of legal tech, but we've got:

- Horizontal positioning (with target verticals) as a broad platform play.
- Vertical positioning, as a classic Vertical SaaS-style play.
- Wedge (or point solution) for a target ICP play.

All based on riding the AI wave and trying to prove out your value in those specific ways. Which is the right approach? Which should you follow?

No easy answers, but a few observations.

Who wins?

Note that Spellbook had, at the time of writing, only raised ~$30M total (last round $20M Series A) vs. $161M total for

Hebbia (last round $130M Series B) and $206M total for Harvey (last round $100M Series C). Different rounds; different stages; different ambitions. But make no mistake — the race was on to be "the co-pilot for X," i.e., the co-pilot for lawyers, and perhaps for knowledge workers more broadly, as Hebbia sees it.

But which company has the right positioning strategy? And when a market is in full land-grab mode, how *should* you position your company?

Well, to turn the question around, the thing to do to understand your positioning opportunity is to understand the market you're in, the resources you need, and how markets — especially technology markets — evolve.

Let's go on another quick detour, this time into the world of positioning as market strategy.

Segment & sell vs. cast & catch

We touched on this earlier, but what separates super positioning from a lot of popular positioning advice is that the classic advice leans pretty heavily on the left-brain laser beam mode of attention. Look down the y-axis to focus narrowly on what you've got *today*, how you compare to others *today*, and who you're the best fit for *today*.

There's good reason for this — buyers are usually buying what you can do for them today, not some fuzzy vision for the future, so it helps to be very clear about that. Sell what's on the back of the truck, as they say.

I call this the *segment & sell* approach.

This is something B2C companies can't do and the B2C brand academics don't understand about B2B. In the B2C world, if you want to be a big brand, you have to use mass advertising. That's your route to market. And that pushes you towards reach — you need more and more reach to build your business. This means mainstream, mass-market products advertised through mainstream, mass-market advertising.

Sometimes, at the SMB end of the B2B world, those tactics can work to build a very strong position in the mind of your buyers, as we'll see in our **own it** chapter.

This is the "cast & catch" approach — you cast a big net and try and catch as many buyers as you can.

So we've got *segment & sell*, which is common for sales-led B2B companies, especially if they go outbound, and *cast & catch* where there's a much more mass advertising, or inbound lead-generation approach.

What should you do when you're trying to ride a wave?

People from the B2C brand world would say you should go for maximum *reach* — sell as much as you can to as many people as you can. That's what big companies do after all — that's why they're big! They advertise broadly in an 'always on' way across as much of the market as they can. They do this to try and build and — more often — sustain brand memories at scale, so that when buyers are ready to buy, they recall their brand and are therefore more likely to choose their chocolate bar, beer, toilet paper, or whatever it might be.

The classic B2B tech advice, however, as popularized by Geoffrey Moore in 1991's *Crossing the Chasm*, the tech industry's original market strategy bible, was all about *niche*. Stay focused, zero in on a segment where you can be a big fish in a small pond, start with the early adopters and 'user innovators' (as we'll touch on in the next chapter), develop unquestioned market leadership in your niche, and then pick off adjacent niches and more mainstream market segments as you grow.

Moore has a whole cinematic universe of metaphors for what can happen as markets develop, including chasms, bowling alleys, tornados, and gorillas, but for now, let's just think about this as our niche segment & sell strategy.

Niche to reach

These two broad approaches represent the *niche to reach* dilemma. Yes, big companies sell more (reach), but small companies have to start somewhere (niche), and ideally over the lifetime of a B2B company you go from one to the other.

But when a market is white hot, with the AI revolution creating brand new market opportunities (no one was talking 'co-pilot for lawyers' or thinking about the 'Excel for knowledge work' previously), again, what's the best approach?

This is where we need to think more along the right-brain x-axis and figure out how our positioning fits in a market that's rapidly developing.

Diffusion

A technology market rapidly developing and maturing is a well-studied phenomenon. What's happening is called "diffusion." That is, an innovation 'diffuses' (spreads) in a market, as Everett Rogers described in his landmark *Diffusion of Innovations*, way back in 1962. That book gave us the classic tech adoption curve of:

- Innovators
- Early adopters
- Early and late majority
- Laggards

When an innovation like AI is diffusing in the legal tech market, or even broader white-collar knowledge-work market, segmentation within that market *only makes sense if that's all your resources allow.*

For example, if you're a small startup, you can only engage so much of the market, so you *have* to be very deliberate about where you apply your effort. That's what Spellbook is doing with their tight positioning, focusing on a 10x solution for a very specific ICP (transactional lawyers).

But when there are insane levels of *pull* from the market and the wave is providing that all-important momentum, then *if you have the resources* it can absolutely make sense to go broader.

If you're VC-backed, the end-point is always as much reach as possible, after all. And when these waves create an opportunity to cast a much bigger net — why *not* try and be the platform for knowledge work, for example? Why not try and be the 'co-pilot for X' for a specific role or industry, where you know the jobs-to-be-done and what LLMs (and perhaps your custom mix of models) can do for them?

For example, the enterprise-focused Glean, raised more than $600M (including a $260M Series E) for its "Work AI platform".

Glean aims to provide a search and automation layer above a company's existing docs and SaaS tools so workers can use AI to pull info from — and push info to — their CRM, their CS tools, or their company knowledge base, for example. "Work AI for all" is their tagline, and that's the very opposite of niche — it's maximum right-brain reach.

Either way, there's so much opportunity for what these strange new co-intelligences can do — and so much demand from the market — that it's worth thinking hard about your vision and how far you could go with foundational models and, say, $100M in your pocket (*bless you!*) to make your vision a reality.

(That doesn't mean maximum reach will always work, of course. Hebbia, as a coda to the description above, has since dialed back their positioning to "The AI platform for finance" rather than all knowledge work or even legal knowledge work. Maybe Spellbook's more orthodox, niche-focused approach will prove correct. We will see!)

Concept creators — CRM

AI wave-riding in B2B is generally about applying this new general-purpose technology to business workflows in new, innovative ways. The power of the AI wave is so strong that it, as we've discussed, does most of the narrative heavy lifting for you, at least as far as the 'why now' question goes. People know what AI is, know they need it, and want to see what you've got.

Eventually, however, categories mature, and it's far more incumbent on your company to come up with a new narrative to explain why your technology is necessary.

Sometimes the companies that pull this off get referred to as category creators, but — to keep flogging a dead equine — I think that's usually a bad metaphor for what's going on.

I like thinking about concepts instead. We will look at some classic literal category creators in a moment, but first let's think about concept creators who take a bite out of an existing category, like CRM, with our neat *concept = new change + new playbook + new tool* formula.

Let's focus on Salesforce, HubSpot, and Drift, who all used

an incredibly strong story 1 narrative to put their new *concept* on prospects' radars in a big way.

Salesforce

The OG of SaaS rode the trend of SaaS itself — well before we called it "Software as a Service" — and started out as an 'inexpensive web-based solution.' Salesforce, who in the dot-com era once famously staged a 'protest' marketing stunt at a competitor's conference with their "No software" campaign (as in, no on-prem software), rode the SaaS wave all the way to their current $200B+ market cap.

To use our **new concept = new change + new playbook + new tool** formula, Salesforce new concept was:

- Ride the new change of *the internet itself.*
- Launch a new playbook for software adoption — no more going through IT, you can sign up on the website (crazy!).
- Pave the way with a new tool that enabled all of this — remember, there's little precedent for this at the time, so they had to be the ones to create it.

Salesforce's journey is indicative of enterprise SaaS as a whole — it started as a plucky underdog with a freemium offer in a 'crowded' category, caught a wave, and was able to grow and grow and grow by selling and selling and selling.

That growth included moves into adjacencies (customer service, marketing, commerce), acquisitions (Tableau, Mulesoft, Slack), and attempts to catch new waves, betting heavily — with everyone else — on AI.

That little wave we call SaaS has taken them a long way, and it started with a vision for buying and using software in a way that seemed completely contrary to the wisdom of the time. I mean, who in their right mind would hand over their precious customer data to some random company on the internet?

Everyone, it turns out.

HubSpot

HubSpot actually started as LegalSpot, a niche solution for legal firms, but they pivoted, went horizontal, and rode the organic digital marketing wave with their "inbound marketing" concept instead. The actual category they would pioneer was marketing automation, before pivoting to CRM. Their market cap, as of writing, is about $30B.

To use our **new concept = new change + new playbook + new tool** formula for HubSpot, HubSpot's new concept of inbound marketing was based on:

- The new change of organic digital marketing — initially blogs, SEO, and social.
- The new playbook for adapting to this change was to adopt an "inbound" marketing approach as opposed to the (supposedly) old "interruptive" approach of advertising and sales outreach.
- The suite of tools HubSpot built helped unify and automate much of the jobs-to-be-done around this work, hence they were seen as a 'marketing automation' platform.

HubSpot gets credited as a category creator, which is true in a *concept* sense, and they're now an incumbent CRM player. But their early success was a product of their vision — their ability to see the organic digital marketing wave, wrap up the change in a catchy narrative and playbook ("inbound marketing"), and repeatedly prove their value with their customers through several pivots and expansions while truly owning their message in the broader market.

In the 2010s, HubSpot was inbound marketing, inbound marketing was HubSpot, and their innovative, partner-driven GTM approach ensured that everyone knew it.

Drift

Founded by ex-HubSpot exec David Cancel in 2015, Drift took an incredibly strategic approach to the CRM market, riding the messaging wave with their "conversational marketing" approach to the use of live chat. They were acquired at a valuation of approximately $1B despite competing in a category where Salesforce had a 16-year head start.

In terms of our **new concept = new change + new playbook + new tool** formula, in Drift's case:

- The new change was messaging (in the chat sense). We're messaging our friends and family all the time. We're using chat for support. Why not chat with businesses when you want to buy, too?
- Drift's playbook for this new era was "conversational marketing" which Drift evangelized far and wide.
- Naturally, Drift provided the freemium tools to get started with this approach, and those tools *had* to be good because competitors quickly started labeling their own stuff "conversational marketing," too. (Narrative windows can close quickly.)

Cancel was, like HubSpot, incredibly deliberate with his strategy. He had the vision — he saw the impact messaging was having in society at large. He had the narrative — he wrapped that change up in a "conversational marketing" narrative Drift could own. And he had the product and execution chops, leading an incredibly effective team to execute on a product (website chat) that had seemingly already been done, especially by companies like Intercom, and used that as a wedge to try and get into the bigger CRM market.

These three examples — Salesforce, HubSpot, and Drift —were all ultimately CRM plays, which goes to show you that wave riding and category creation aren't the exact same thing, given CRM was already an existing category. You don't *have* to invent a category — or even be first in it — to achieve an incredible exit, and indeed most startups don't.

Instead, one option is to ride a wave and take a very healthy bite out of an existing category. Salesforce thought they were entering a crowded market when they launched in 1999. HubSpot must have thought the same when they pivoted to CRM. And Drift entered the CRM market precisely *because* it was a crowded market, in a swim-for-the-red-ocean sense.

Big markets signal big demand, after all, and the CRM market continues to see innovative new players emerge and shake things up. The Notion-esque CRM Attio, for example, which has raised $56M over 2023 and 2024, is aiming to disrupt Salesforce, once the disruptee, as the startup circle of life of incumbents and disruptors continues unabated.

Category creators

AI-first startups have the luxury of insane demand for AI solutions — businesses know they need to put AI to work.

Concept creators riding a wave *within* a category can also rely on the fact that there's strong, established demand for the parent category. Businesses know they need a CRM, for example.

Some startups, however, need to create — or at least pioneer — the new category itself. This is still about a new concept targeting right-brain attention per our change + playbook + tool formula, but the result really is a new category of software. (Category creation does happen!)

This is a high-risk, high-reward strategy — you have to cultivate demand from scratch, but that blue ocean (so to speak) of opportunity can prove very lucrative indeed.

Here's a quick look at the classic category creation examples of Gainsight (customer success), Yammer (enterprise social networking), and Qualtrics (experience management).

Gainsight

Gainsight hitched their wagon to the emergence of customer success as a discipline and are credited as the customer success 'category creators.' In reality, they didn't create the "customer success" category from scratch, but they did start riding the wave early and rode it all the way to a juicy $1.1B acquisition.

In change + playbook + tool terms, it was the SaaS wave itself that brought about the need for 'customer success'. Mere customer *support* was no longer enough — customers needed proactive help to *succeed* with the tool, not just be reactively supported if and when they asked for it.

(Another reminder that in B2B, customers aren't buying tools *per se*, they're buying outcomes, and they often need, and will pay for, a lot of handholding to achieve those outcomes!)

SaaS companies themselves needed customers to succeed so they wouldn't churn. The playbook, then, was a focus on churn. Having a specific metric to target is a very powerful thing, and

Gainsight's tools could, as their early homepage back in 2013 described it, "Reduce Churn, Increase Up-Sell and Drive Customer Success." The change, playbook, and tool all went hand-in-hand.

Yammer

The OG category creator, Yammer, created the category "enterprise social networking," or ESN. This was a true category in the literal category-of-software sense, a category that's still with us today. But it wasn't the category *name* that particularly mattered — they just put "enterprise" in front of "social networking" — it was the wave that mattered.

Founded by David Sacks (yes, of *All-In Podcast* fame) back in 2008, Yammer built social networking-style tools for the enterprise and, crucially, combined it with a 'bottoms up' (i.e., product-led) GTM motion *and* a sales force that eventually outcompeted their main competitor, Jive.

Founders can get hung up on the name for the new category or concept they're trying to pioneer, but what's interesting about Yammer is that "enterprise social networking" was self-explanatory. It provided instant clarity on how Yammer was different from the competition. It was social networking... for the enterprise.

In the change + playbook + tool sense, the broader social wave was of course B2C social networking (with Facebook and Twitter blowing up at the time), the playbook was bringing this new UX to the enterprise, and the tool was a matter of building a platform to match. That might sound like an obvious opportunity in hindsight, but the dominant social paradigm at the time was, as Sacks tells it, blogs and wikis, and Yammer had to forge their own new category to distinguish themselves from the 'old school' players they were competing against.

(The other lesson from Yammer is that PLG is great for building bottoms-up momentum, but you still need a sales team to close enterprise deals, a lesson startup founders will keep relearning until, one imagines, the end of time.)

Qualtrics

Qualtrics famously coined the term "XM," or experience management, to help them *transcend* their software category of survey software. And it worked! But it happened when they were a very mature company, having spent more than a decade and a half riding the digital transformation wave, which put them in a position to launch a new concept, playbook, and brand.

The change was the "experience economy," the playbook was experience management (XM), and the tool was their mature survey and analysis platform rebranded for a new era.

What's remarkable about Qualtrics is the extent to which they were able to sell their *own narrative* of being a category creator. There was no AI-like wave to take them there; they did it themselves, and they started in an incredibly narrow niche, which brings us to the next super positioning strategy, 'find it'.

Your turn — what's your wave?

- What wave (if any) is driving demand in the segment you're targeting? (If you answered "AI" — and I know 9 out of 10 of you did — think about what the wave is about. Is it specifically about automation, analysis, conversation, acting as a co-pilot, or something else?)
- Are you trying to ride that wave to create a new category, transcend a category, or carve out a new position in an existing parent category (like CRM)?
- Are you going for the horizontal platform play (Hebbia), the strict vertical play (Harvey), or the 10x point solution for a specific ICP (Spellbook)?
- What role does external change play in your current narrative or messaging? What role *could* it play?
- What does your *concept = change + playbook + tool* formula look like — especially the playbook part? What can you learn from the CRM players & category creators in terms of owning a playbook?
- Finally, how are you proving out the new value for a given set of jobs-to-be-done that this wave creates? Your ability to prove that value is what will make or break your startup.

Next, let's look at finding value in a niche.

12

Find it & discover unique value in a niche

Find it in a nutshell:

- **Goal:** Find a way to be **super useful** to your audience.
- **Strategy:** Find your **unique** value by looking down for a specific segment to exploit, be that a vertical, job-to-be-done, user innovation, attribute, niche, or 'best customer' ICP.
- **Attention:** Active left-brain (looking down).
- **Action:** Find it — either do the analysis to find the niche yourself or do the work to see what your most innovative customers are doing.
- **Position:** Your tool or innovation is a revelation to a defined set of folks who really need it and very quickly *get it*.

You don't have to deliberately build your positioning around a new wave to build an incredible B2B startup — we're all riding the digital transformation wave after all. Sometimes, it's about drilling down into (or stumbling across) a niche and building the right tool for the right people at the right time.

This is the world of more left-brain positioning by analysis — finding who you're the best match for and then zeroing in on those best-fit customers.

But why do this? Why limit yourself to just one niche? Why not just sell to whoever wants to buy?

There's some crucial logic at work here that comes from classic tech market strategy (tip o' the hat again to Geoffrey Moore and *Crossing the Chasm*).

Niche momentum

Because you're a small startup and no one knows who you are, you can't afford to pay for every marketing touch for every sale you want to make. Instead, you need some degree of word of mouth, or at least reference-ability.

That is, you need people telling (or tweeting) other like-minded people about your great innovation. The only way they can do that is if they're in the same proverbial room (industry, role, geographic location, Slack group, LinkedIn bubble, etc.) as the people you need them to talk to.

What's more, when your go-to-market strategy relies on sales, and therefore *reference customers*, this niche approach is even more important. When company X comes to buy, and they see that company Y *who is just like them* is already a customer, they're going to feel more confident taking a risk on buying you, especially if you're a startup with mostly unproven or still rough-around-the-edges technology.

ICP as regression

Dave Kellogg has a very pithy way of putting this, saying:

> *While [your] ICP starts as an aspiration, over time it turns into a regression.*

That is, you start out with some theories about who will get the most value out of your product, but after you've been in the market for some time, you need to go back and do the analysis to see who *really* gets the most out of your product.

That helps you refine your ICP (ideal customer profile), and therefore the way you position around, and sell to, them. This is explore/exploit, in other words.

This approach works best when you're doing 'segment & sell' (as we discussed last chapter) because you now know who you need to go and sell to. Strategy is choice, and choosing your target customer to position around is one of the most important choices you can make.

Hand-held sales

One of the best founders to pull this off and document the process was Nathan Barry of ConvertKit.

Barry bootstrapped an email marketing startup in the face of overwhelming competition from Mailchimp *et al.* by being laser-focused on bloggers, doing direct outreach and sales to them, hand-holding them through migration, and picking off reference customers in each micro niche one at a time to build his startup brick by brick.

Eventually, the flywheel kicked in, word of mouth started to build, and the company now generates tens of millions of dollars in revenue every year, fully bootstrapped.

Vertical SaaS

At the venture-backed end of the SaaS spectrum, 'vertical SaaS' has become a popular strategy because it doubles down on this build-momentum-in-a-niche approach. However, the 'niche' in this approach can often be an entire industry.

Famously, Veeva, who thought they would start in the pharmaceutical niche before moving on to a bigger niche, found it so profitable that they ended up building a public, ~$30B business in their original vertical and now position themselves as "The Industry Cloud for Life Sciences."

Whether bootstrapped or venture-backed, the logic is the same. Position in a niche, sell *only* to a very specific kind of customer, get the flywheel going, and watch that compounding momentum drive your growth.

Proactive vs reactive

There are, broadly speaking, two ways to find your niche:

1. **Proactive — pick your niche/vertical upfront:** You see a problem, build a tool, and take a very deliberate, founder-led sales approach to selling to (and onboarding) customers by hand, one by one, building with them as you grow. Alex Kracov, the founder of Dock, is another great example of someone doing this and documenting the process.
2. **Reactive — pick your best customers once in the market:** Plenty of folks build something, start selling it, and are generally happy to sell to whoever's buying. If this works, great, but in the long run, this can fail to generate that all-important niche momentum. The word-of-mouth fire never gets lit and growth never really takes off because your customers aren't in the same 'room.' In this case, you need to analyze your customer base, find your best customers, and try to build momentum that way.

Some leadership teams will forensically analyze *dozens* of potential niches and adjacencies to see what segment appears the most attractive.

Others just YOLO it and follow their nose.

And others study their best customers. This approach is something that the queen of B2B positioning (and massive inspiration), April Dunford, strongly advocates. Even better, Dunford gives us a handy methodology for doing this kind of positioning, which we'll touch on here.

(We'll only touch on it because Dunford's work is so popular and her efforts have been so tireless that there's little point in rehashing it here. But again, let me emphasize how much I respect and admire everything Dunford has done, from the incredibly hard slog of trying to nail down a concise (!) methodology to the even harder slog of getting it out there through sheer force of will. A true legend of the marketing world!)

The second prong

This is where the second prong of super positioning comes into play — competitive positioning.

Here, we're particularly interested in how you position your product closer to demand (customer value and jobs-to-be-done) and away from competitors (through uniqueness and differentiation), making your insane usefulness *for a given segment* clear in the process.

That's why, in Dunford's approach, for example, positioning starts with competitive alternatives and goes from there.

Starting with JTBD

Interestingly, Dunford's approach to competitive alternatives is informed by the Jobs to be Done framework we discussed in the previous chapter, and we'll explore some more here. As Dunford discussed on *Lenny's Podcast* in 2024:

> *Eventually, Clayton Christensen solved this problem for me. I was reading everything I could get my hands on about the Jobs to Be Done (JTBD) theory, and I had the realization that the starting point had to be competitive alternatives. If it wasn't, what we ended up with was positioning that sounded good in the office but didn't work with customers because it wasn't differentiated.*

By competitive alternatives, Dunford means competing ways to do the job-to-be-done. That's what she argues grounds positioning in the customer's world.

That is, you're not looking just to differentiate yourself from competitors over some new feature that customers may not know or understand, even though it might 'sound good in the office.' You're differentiating yourself from the current or potential way a customer actually gets a job done, which might include manual processes, spreadsheets, general-purpose software, and so on.

That's the measure of usefulness we care about.

So while Dunford's positioning approach starts with competitive alternatives, we could also say it starts with the job-to-be-done. (Indeed, a lot of JTBD case studies look like positioning work to your humble author!)

That's a very useful insight because it means we can bring all the JTBD literature to bear on our competitive positioning problems, especially when we're trying to find unique value in a niche.

Everything in B2B boils down to a job-to-be-done in the end!

Emotion-driven positioning

We're going to talk about tables and spreadsheets and analyzing jobs-to-be-done in a moment— all very left-brain laser-beam kinds of activities. But first, let's note the role emotion plays in positioning. I don't mean that in some gooey, how-I-feel-about-a-brand sense. I mean that in the sense of *what pisses you off*.

Dunford says positioning starts with competitive alternatives, and eagle-eyed readers will note that starting with competitive alternatives (category creation) is how I frame the cover of this book.

That is, I started from the job-to-be-done (find a dominant positioning strategy for startups), identified the existing alternative ("category creation"), and positioned against that. Why? It pissed me off!

The truth is, however, that the way I got there wasn't through careful analysis; it was vibes. I was trying to find a hook because I was failing my own **hook test**. I wish I was analytical enough to figure that out ahead of time, but the reality is I was just noodling away until I felt like I had something strong enough to put on folks' right-brain radars.

It turned out that the *emotional energy* of what annoyed me was a bigger driver for my competitive positioning than pure dispassionate analysis. Maybe that will be true for you, too.

After all, if you're going to spend a lot of energy trying to build a position in the collective mind of the market, that energy has to come from somewhere, and an emotional driver is as good as any!

Pain = motivator

This position-against-things-that-piss-you-off approach seemed to be at least a little bit true for Dunford, as well.

Dunford's competitive foil was the classic *positioning statement*. You know the one, the madlibs style "We're an X for Y, and unlike Z, we offer A, B, and C."

Dunford really, *really*, hates those. On her site she writes how they "suck", how they're "a candidate for most useless business tool ever," and how they're "harmful" and you "should never do one."

That's some emotion-driven energy! I love it.

Pain is life's great motivator, and frustration at how everyone is Doing It Wrong is one big motivating pain.

What pisses you off about the way things are currently done in your industry or market?

Outsized ICP value

Pain aside, the alternative to the positioning statement is, for Dunford (and I'm brutally simplifying here), a methodology and a table for analysis.

To paraphrase the methodology:

1. First, we identify competitive alternatives (in the jobs-to-be-done sense, i.e., status quo alternatives, not just peer competitors).
2. We then identify what makes you unique, in comparison, in terms of features, benefits, and the value you offer, and roll those into value themes.
3. Finally, we figure out who cares a lot about that value, i.e., your best customers.

And then we try to sell just to those people.

That is, you calibrate what makes you unique (differentiation) relative to all alternate options with what makes you most valuable (demand) for a specific set of customers, and you build your ideal customer profile (ICP) where those things intersect.

Or, if I can simplify further, we might look for existing customers where:

- **Your differentiation + their specific demand = insane ICP usefulness.**

That's how you win deals over alternatives.

"Oh, you have this specific need? Well, we have this specific solution that creates this specific outsized value just for you."

Or, if ICP value is similar, your 'best customers' might be those who pay you the most, or require the least support, too. Either way, you drive go-to-market (GTM) efficiency by making sales easier, more profitable, or both.

Write. It. Down.

How do you apply this to your situation? Time to crack open another spreadsheet or doc and start writing stuff down!

You can call this a one-pager, canvas, table, spreadsheet, whatever. Even if you choose my preferred narrative approach (**chapter 9**), the point is to make sure you actually *write your positioning down.*

That is, there's no super-secret special sauce in any particular template. The point is to *get writing* and do the actual analysis.

Get listing

This might seem obvious to marketers — we tend to be the writers, after all — but you'd be amazed at the number of sales-driven orgs where the culture is far more verbal, and they can go for years without ever really writing their positioning or ICP down in a concrete way.

But write what down? Dunford's approach is, as mentioned, to list your features, describe their benefits, list the customer value of those benefits, and then group your value themes together.

This approach is great when:

1. You want to rigorously define why you're better than the

status quo (manual processes, spreadsheets, etc).
2. You want to rigorously define why you're better than peer competitors.

This is an approach born out of competitive positioning in enterprise sales in particular. That's where the customer's laser beam really comes out and they'll go over your features, why you're different, and why they should care, in forensic detail.

Your JTBD canvas

Personally, I like to go a bit further and come back to Jobs to be Done.

Over the years, I've seen a bunch of frameworks put together by consultants where, if you rip the mask off Scooby-Doo style, underneath you'll find it was Jobs to be Done all along.

And I would've gotten away with it, too, if it weren't for you meddling kids!

(Tongue is firmly in cheek here, and I say that about myself as much as anyone. Obviously, people bringing their own experience and perspective to the theory has value, and hopefully that goes for this book, too.)

If Jobs is where it starts, I find it helps to go back to the OG of JTBD, Tony Ulwick, and go from there. I like to riff on his definition of Jobs, which I paraphrase as:

1. Customers have needs; those needs are a function of *jobs* they want to *get done*.
2. Jobs are *stable over time, solution agnostic,* and ideally expressed in a brief statement.
3. Customers buy products to help them get their jobs done, ideally on one platform.
4. Customers can express on what *dimensions* they'd like to improve how they do those jobs — faster, cheaper, more often, less often, and so on.

Your turn

With that definition in mind, I strongly encourage you to make your own spreadsheet or table to think through how this applies in your case, listing out:

- **Role:** The customer or role, i.e., the person doing the job.
- **Job:** A given job they want to get done (don't assume the solution here, keep it focused on the problem).
- **Status quo:** How they currently solve it.
- **Features:** Your feature/s that help get that job done.
- **Outcomes:** The dimensions of improvement your features enable in a very quantifiable way.
- **Differentiators:** How your approach compares to the competition.
- **So that:** The big picture outcomes that are less quantifiable but are the big emotional payoff — get your day back, stress less, feel in control, feel in command, be dramatically more productive, and so on.

The accuracy of your version of this table — and the insight it contains — is a function of your ability to look *down* into your customers' lives.

And not just what you've seen personally, but what your entire company has seen in terms of case studies, outcomes, customer research, product development, and so on. Everything you do as a company, in a sense, is working towards that outcome of a job solved in a particular, differentiated way that makes you unique.

Sugar-coated broccoli

Having the language of Jobs (along with features, benefits, and outcomes, for example) is very helpful — we need universal ways to describe needs and outcomes — but sometimes the real positioning value is using JTBD to find **message/market fit** for your startup.

Let's say you're a very horizontal tool like Notion, for example, with a big, whole-new-way-to-work pitch. But is that why they sign up and buy?

Here, it pays to be laser-focused on the jobs people are actually getting done with your platform, and to reflect that language back to prospects so they get it. They simply won't know or care otherwise — I mean, how could they?

Before their AI push, for example, Notion's homepage did this very directly by highlighting three main areas — wiki, docs, and projects.

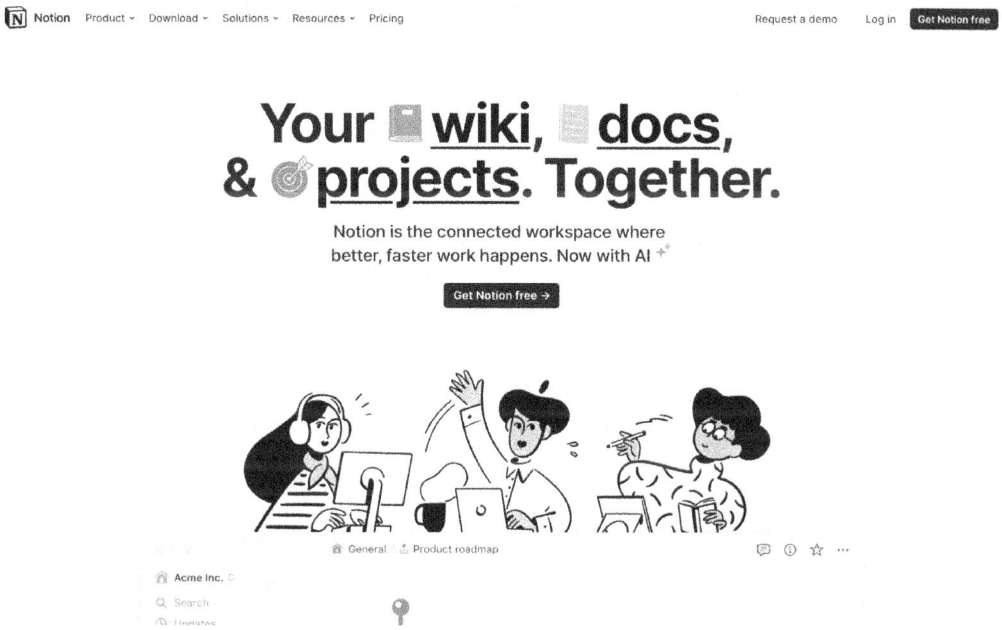

Call them use cases, call them jobs, call them whatever you like, the point is that the language matched how prospects thought about the actual tools they needed.

Contrast that with Notion's big right-brain vision, which was, according to founder Ivan Zhao in an interview on Lenny's Podcast, that "everybody can create their own software."

The problem was that nobody cared.

Notion tried, in the early days, to let people build with the tool, but they just didn't care. So instead, they said (edited for clarity):

> *"Let's hide our vision (which is that everybody can create their own software) in the form factor that people do care about. What kind of tool do people use every day? Productivity software. [...]*

> *"The world is not like you. The world [does not have a] developer or designer's mind. The world only cares about what's in front of them."*

So they hide their vision instead:

> *We call it sugar-coated broccoli. People don't want to eat the broccoli, but people like sugar, so give them the sugar and then hide the broccoli inside of it.*

That's a great turn of phrase, and the net result is putting the jobs people actually want to get done front and center.

I like to think of it as message/market fit, and of course it goes deeper than just the literal message in your homepage hero, it goes to how you frame your product, the templates and integrations you offer, the case studies you highlight, and so on. But it ultimately matches what's "in front of" the user, and that's their jobs-to-be-done.

Differentiation vs. fit

When we talk about competitive positioning, most of the time that means highlighting your differentiation — how you're unique relative to the competition. That's because most B2B tech products and platforms, especially in SaaS, are, let's face it, pretty same-y.

But let's remember our definition of positioning — it's the act of being as far from your competitors as possible *while being as close to your customers* as possible.

Notion's problem, initially, was that their distance from their competitors was *also* their distance from customers, and that doesn't work. In that sense, Notion was *too* unique, too 'out there' relative to what customers wanted in their day-to-day. They had the differentiation, but they didn't have the fit.

With a combination of right *and* left-brain attention, however, we get both the big vision and unique product *and* the focused specifics that people care about in the day-to-day.

That requires the kind of intense left-brain focus we're

interested in here — focusing down the y-axis on the actual jobs-to-be-done people can achieve with your product, and the language they use to describe those jobs.

For Notion, this focus — and all the templates that followed — helped propel them into becoming the juggernaut they are today.

Honestly, I've seen the concept of Jobs to be Done blow teams' minds — it *finally* gives them the language to articulate the needs they've been selling against and building for over all these years.

We understand the world through metaphor, and 'the job to be done' is a very powerful metaphor in tech.

The L in R/L/R

There are limitations to this approach, especially when it comes to your initial positioning. You still need a big idea to put on folks' radar in the first place and that's not always found at the bottom of a spreadsheet.

When analysis (which is always backward looking) is your dominant mode of how you do positioning, there's often a bit of handwaving about your initial positioning. You 'just' put your product out there, see who buys it, and *then* figure out who your best customers are.

But that initial net you're casting still determines who the 'best' customers are that you can pick from. There's a lot riding on that initial positioning!

For example, In a 2024 podcast interview with Brian Rhea, Dunford said:

> *The best time to actually look at your positioning and think about tightening it up is when you have actually been in the market for some time and you have got some traction, to the point where you can start seeing the patterns in, "These are the sorts of people that love us. And this is why." [...]*
>
> *So, you're fine to go out to market with your positioning theory but be prepared that, in almost every case, the positioning ends up changing as you start getting more experience with that market and you see what works and*

> *what doesn't.*
>
> *In fact, most of the time I counsel early-stage startups to keep their positioning a little bit loose for that reason. Because you actually want to get exposure to a wide swath of customers, so that you can then figure out, "Well, who really loves my stuff, and why?"*

This is, of course, all very sensible advice for once you have got some traction and want to know where to double down. It's also very true that early-stage startups pivot and reposition all the time, so you shouldn't get too wedded to your initial approach. Nor should you try and predict everything upfront, which is a recipe for overthinking and second-guessing yourself. (Not that your humble author would know...)

You need to keep sampling reality — customer reality — and learn as you go. It's action, not endless speculating, that ultimately generates the information you need.

But, again, how do you get that initial traction, and what kind of traction is it?

You have to be thinking hard about what you're putting on folks' right-brain radar before you get out the laser beam and start doing the analysis. Indeed, your initial idea has to be strong enough to create some signal you can find *through* this left-brain analysis.

You ideally want to be spoiled for choice when it comes to finding your best customers, not scratching around trying to find some kind of hook from a bunch of so-so deals, and this can throw folks off.

Obvious vs. fuzzy

Initial positioning aside, what are you actually looking for when it comes time to deploy your left-brain laser beam? What are you hoping to find in the middle of this R/L/R jump?

Here, simply knowing what to look for can go a long way, especially when a niche positioning opportunity slaps you in the face.

This is especially true when you see what unexpected and innovative things customers do with your product, essentially finding your niche position for you. (Which, again, isn't necessarily found in a spreadsheet of features and imagined outcomes.)

Other times, however, it's more fuzzy, and a niche isn't so clear-cut, and you instead need to camp out on a specific *attribute* in the face of intense market competition.

Let's touch on both.

Obvious — user innovators

If you're trying to find a new positioning opportunity, one of the best questions to ask is "Are any of our users doing something cool (or crazy) with our product?"

Sometimes, very obvious positioning opportunities present themselves when the current way just isn't working. Perhaps you're fortunate enough to spot a handful of users — or even just one — doing something innovative with your product, and the positioning lightbulb goes off.

That was true for Loom, the video recording software, as we'll see in our examples below.

This is the world of *user innovation*, and it's gold for positioning. Why? Well, it turns out R&D is very expensive for startups and companies alike because trial and error is very slow. A startup might build a product and take a shot at positioning it, but from their point of view, that's just based on the trial and error they could do internally. And that might not be much!

Therefore, if we think from first principles in terms of vision, the opportunities for a product a startup can see, in a very literal sense, are inherently limited. There are only so many eyes doing the looking.

But if you put that product in the hands of dozens, hundreds, or thousands of customers, suddenly you have an infinitely faster R&D trial-and-error flywheel, at least in terms of uncovering good-fit jobs-to-be-done. There's just much more 'looking' going on.

This makes it a very powerful tool — you're selecting the best uses of your product from many trials, but also the least reproducible because, other than making sure you're keeping an eye out for novel uses, it's dependent on what your user innovators come up with.

There has to be some value there to begin with for people to adopt the product, but even companies like Apple with their we-know-what's-best culture rely on this approach to find the best use cases for their products like Apple Watch (notifications and health tracking) and Apple Vision Pro (movies and virtual screens for your Mac).

This phenomenon of *user innovators* has been studied for decades by Professor Eric von Hippel. It's amazing what research has established that we occasionally rediscover in tech (which, for me, came via Professor Ethan Mollick, one of the most prominent voices in AI.)

This kind of user innovation can be an absolute no-brainer to go after, especially if your initial positioning flops hard. But you have to do the work to find it. By definition, these are edge cases that you have to tease out through interviews and analysis, and just by being very receptive to customers doing weird things.

But if whatever you're doing now isn't working, why not try and find what is amongst the people who are already using your product?

Fuzzy — unique attributes

Sometimes things are working, or working well enough that you don't want to risk a massive pivot to something else, but you know your positioning could be tighter.

Here, especially in competitive markets, positioning opportunities might not be hit-you-in-the-face-obvious, but you can still find an attribute to position around, something to be 'more about' than the competition, and if that keeps you competitive in a growing market, you can grow *with* the market.

A client of mine spotted a glaring gap in a customer workflow while on-site with the customer and built automation functionality to close that gap. That gave them an attribute — automation — to be 'more about.' That gave them an angle to be super useful and position around, as we'll explore below.

Positioning and *gestalt*

I describe this as "fuzzy" positioning but what I really mean is we're looking for a *gestalt*, which is a German word for "an organized whole that is perceived as more than the sum of its parts."

It's about a unique configuration that's driven by positioning-by-attribute. It's inherently more varied simply because there are many attributes you can be 'more about' than the competition, and the *gestalt* you create is unique to you.

If you pick your attributes wisely, however, you can find a compounding position and carve out a unique space both in your market and the customer's mind, perhaps through positioning around:

- **Price & segment**: That is, targeting the low or high end of the market. (Workbooks in CRM, which positions itself against Salesforce's pricing with lines like "No sales tactics. No overpricing. No bullsh*t.")
- **Point of view & UX:** A specific point-of-view on the market, perhaps with a UX to match. (Linear, as we'll see next chapter.)
- **Wedge attributes:** Usually thanks to feature areas they've invested in, especially when leaning hard into a point of view, perhaps around automation, collaboration, intelligence, and so on, or just doing a specific JTBD 10x better. (Spellbook, as we saw in the previous chapter.)
- **Niche category:** A *narrow* sub-category they've created — for instance, they're a particular *kind* of CRM or database, one for product-led sales (Pocus). They're not creating an all-new category — everyone knows it's still a CRM or a database — but they've found a new, innovative niche and a fresh way to map the parent category.
- **Competitor contrast:** Sometimes you can call out an incumbent directly, for example, Fathom Analytics calls itself "A Google Analytics alternative that's simple & privacy-first."

(Note their privacy-first attribute comes from a bit of wave riding, too, specifically the GDPR legislation & resulting need for compliance!)

The point of a *gestalt* is to roll these together in a compelling way.

Maybe you're the cheaper version of an incumbent with a better UI derived from a specific point of view about how to do a specific job or complete a workflow.

Or maybe you're the developer-focused product analytics platform PostHog, with a homepage that's as thorough as it is unhinged, featuring, among other things, an illustrated section with their cute mascot in a Godzilla costume terrorizing a city and a boxed-software call-to-action helpfully framed with a giant header that says "This is the call to action."

But PostHog also competes aggressively on price, claiming to have the "lowest pricing for every product" and to be "always cheaper than the cheapest major competitor."

That's a unique combination of attributes!

Again, building a super position is about compounding specifics — the unique aspects of your product and company that make you, you.

Refining vs. reinventing

One of the things that took me a long time to understand about positioning in the *Obviously Awesome* sense was that the success stories — ones where a company seemed to reverse their fortunes through positioning alone — tended to be the rare cases where an obvious user innovation was discovered.

That was in contrast to some of the fuzzier "point of view" positioning which seemed more about *refining* rather than *reinventing* your positioning.

What's appropriate for you depends on your trajectory and goals.

The Browser Company, who made the Arc browser, for example, found a niche with power users who loved the product. They'd found their 'best customers,' in a sense — a passionate and vocal group who loved the product. But The Browser Company had raised $128M and aspired to something bigger.

Despite, as they wrote in a newsletter, the fact that a "lot of people loved Arc," the reality was that "for most people, Arc was simply too different, with too many new things to learn, for too little reward." It was more like a "highly specialized professional tool (like a video editor) than a mass-market consumer product, which we aspired to be closer to."

So they went back to the drawing board, released Dia, a new, AI-first take on the browser, and tried to ride the AI wave to create, in their words, "the successor to the web browser" and "have an impact for people at real scale."

That is, they went back to that first R in the R/L/R jump and rethought what they were putting on folks' radars to begin with.

That's a massive reinvention.

If they were just an indie product, however, then refinement would have made sense. They could have doubled down on the power-user positioning and dined out on that, refining their positioning based on who loves them and why.

But that wasn't their ambition, and if you're a venture-backed startup, your positioning ultimately *has* to be a function of your ambition, which that requires a lot of focus on the first R in the R/L/R jump.

(And in The Browser Company's case, resulted in a $610 million acquisition by Atlassian. TBC didn't end up building a super position, but their ambition means they successfully exited to help Atlassian maintain theirs!)

Positioning failure

Going after a niche is common advice for a reason. There's a certain relief and freedom in not trying to compete with everyone over everything — but it takes a lot of discipline to get it right and avoid common failure states. Let's run through a few of them.

Not actually niching

Some folks do positioning exercises, feel good about it, and then... not much happens. That's because they didn't drill down hard enough to focus on a niche (and particular problem) where they could innovate and generate that all-important momentum.

That is, they didn't actually choose a meaningful position; they just kept drifting along with a slight tweak here or there or a lightly refreshed homepage.

Remember, as much as I love **strategic messaging**, copy tweaks rarely move the needle on their own. Positioning requires going back to the **strategic specifics** that are unique to you.

Not getting out of your niche

You sell to early adopters, and they love it. Great! Then you try to cross over to the mainstream (per *Crossing the Chasm*) and discover there is indeed a chasm there waiting to gobble you up. Not so great! OG smartwatch maker Pebble had the misfortune of experiencing this first-hand in the mid-2010s. They made a watch for nerds distributed to nerds through crowdfunding, tried to go after the mainstream market instead, but the product flopped, and they went bankrupt in 2016. (Pebble's founder managed to revive the tech and brand in 2025, however!) Breaking out of an early-adopter niche can be *brutal*.

Staying too niche

But break out of that initial niche you must, and, eventually, contra the saying, you kind of *can* be all things to all people, or at least more things to more people. That's 'how brands grow' (to borrow a phrase, as we'll look at next chapter), and it's why Salesforce, the granddaddy of SaaS, isn't a plucky PLG app anymore but a giant suite of products and sub-brands. If you stay in your niche forever, the world might pass you by. And ConvertKit, for example, is now riding the 'creator economy' wave — even as a bootstrapped company, they didn't stay niche forever.

This is where competitive positioning and the art of more right-brain strategy intersect. You've got to know where you are in terms of your growth and market penetration. Positioning evolves over time, and this is where your own right-brain and left-brain attention comes into play.

Your right-brain radar attention is great for studying your own change-over-time narrative, how the market is unfolding, and how you're making progress toward your greater vision, as we saw in the previous chapter.

Meanwhile, your left-brain laser-beam attention is great for looking down at where your competitive positioning needs to be *today*, right now, for the buyers who want to buy today and not in six months or a year from now (as April Dunford often says).

When these two modes of attention come together and result in a memory association that sticks, you create a super position. And those positions are often found, hence this strategy of *finding it*.

Niche positioning — examples

Let's wrap up this chapter with some examples, starting with Qualtrics, whom we touched on when discussing **wave riding**, and then Loom and my LMS client.

Qualtrics — niche ICP

Qualtrics has been sold and acquired a couple of times for billions of dollars, but the truth is they bootstrapped for much of their existence, initially selling just to *marketing professors* because one of the founders, Scott Smith, was a marketing professor, and they couldn't sell to anyone else! When we talk about getting close to an ICP, that's what it can look like — being super useful for one incredibly narrow role.

But that worked, and eventually, they could sell to business schools, then universities, then finally the mainstream corporate market, in true *niche to reach* fashion of expanding concentric circles.

Loom — user innovation

Loom, the async video software service, originally started as a user testing platform, where their customers could record user tests. But then the team noticed *one user* doing something unexpected — they recorded a short video of *themselves* explaining the *results* of a test to share with their team.

Aha — collaborative async video! That turned out to be super useful, so the team pivoted, positioned around async video with almost flawless product execution to get them going, and was eventually acquired by Atlassian for just shy of $1B.

Spotting one innovative user can have a profound impact on a startup's trajectory!

LMS provider — owned attribute

As mentioned, I helped a past client of mine clarify their positioning around *automation*. They, along with hundreds of other companies, were in the highly competitive LMS (Learning Management System) space. Other companies positioned around *micro-learning, AI-powered learning, collaborative learning,* and more.

Ultimately, my client was acquired by one of their well-funded competitors. Most startup exits are acquisitions, and having a clear attribute you 'own' (and have built out) can help make your value clear to both current prospects and future acquirers.

vSaaS

As we discussed earlier in the chapter, there's also the world of **vertical SaaS** where, like Veeva, you position exclusively in one industry.

For example, construction management SaaS Procore carved out a ~$10B market cap in the construction industry, and restaurant management platform Toast, despite a rocky few years, is worth ~$26B. Much has been written on vertical SaaS over the years, particularly by investors who obviously find those sorts of outcomes rather attractive.

Next, we'll look at our third strategy, *own it*.

Your turn — what's your unique value?

- What pisses you off about the way things are currently done in your niche or industry?
- What unique value have you found in the segment you're targeting? How strong is your vision in terms of being able to look down into the workflows and struggles of the folks in that segment?
- How did you originally see your unique value? Was it through your own lived experience, observation of your customers' problems (like my LMS client who was on-site with a customer), or observation of your customers' innovations (like Loom and video sharing)?
- What niche (if any) are you targeting in terms of a particular role or vertical? Has that made sales easier? *Could* it make sales easier?
- What attribute (if any) are you trying to own so you can be 'more about' something than your competitors and build a unique *gestalt*?
- What role does your unique value play in your current sales narrative or messaging? If you surveyed customers, would they agree that's why they bought you?

13

Own it & associate value with your brand

Own it in a nutshell:

- **Goal:** Find a way to be **super memorable** to your audience.
- **Strategy:** Build your brand through **associated** value in the form of a tight name/need memory association.
- **Attention:** Passive right-brain (i.e., open but not currently in-market).
- **Action:** Own it — you need both consistency and frequency in broadcasting a message you own.
- **Position:** When a buyer needs X, they think Y — your brand. That's the position you own in the buyer's *memory.*

The third prong of super positioning is brand. This is mostly the domain of the B2C world where the marketing science gurus of the Ehrenberg-Bass Institute's (EBI), led by Professor Byron Sharp, have enormous influence.

Their view, based on big B2C brands, is that positioning is inherently a mistake and not how you grow!

These folks tend to study mass market advertising for consumer commodities (including B2B commodities, to be fair), not innovative products that startup and technology companies are trying to 'diffuse' into the market. (Diffusion science has been popular since the 1960s, as we discussed when looking at waves.)

But because super positioning is about the synthesis of seemingly incompatible views and solving positioning's elephant problem ('It's a tree trunk!' 'No, it's a rope!' 'No, it's a spear!'), we can still learn from the brand folks when they're right and ignore their blind spots when they're wrong.

And in terms of being right, they have an incredibly helpful take on what 'brand' actually is, and that's *memories*, as we've been discussing.

Specifically, it's a name/need association that appears first when people search their memories before they search Google. (Brand search has to come from somewhere, right?)

This matters because it clarifies what brand *is not*, too.

Your brand is not just your logo. It's not any elaborate brand "meaning" or (rabbit hole warning) "purpose" some marketing folks might get carried away dreaming up, either. And it's certainly not your sassy best friend or some object of love and affection.

Instead, it's a simple memory/need association. Need drink; think Coke. The value of Coke as a brand has been built over decades, both as a memory association in our minds and in the physical positions they've built in just about every nook and cranny across the face of the earth.

That is, to use the EBI's terms, *mental availability* on one hand — lots of people associate Coke with a refreshing and tasty drink — and *physical availability* on the other, with Coke being distributed virtually everywhere. This approach makes it very easy to buy when your memory reminds you it's a possible solution to the thirst you're feeling.

Coke's brand, of course, is built on its 'distinctive assets,' to use a term Professor Jenni Romaniuk of the EBI has coined. Those brand assets include its logo, its use of red, and its can and bottle design. They're the memory glue between the ads we see and the product we look for.

Brand as experience

Distinctive brand assets might help us connect advertising with a product through its brand, but why do we remember some brands and not others? Why do we end up associating certain products, people, or companies with certain needs in our minds?

Brands are built on memories, but the driver of memory is experience.

In a first-principles sense, we have to have an experience strong enough to make an impression — to catch our attention *and* create a meaningful memory association.

For food and beverage, this is relatively straightforward: just show the product and/or its enjoyment in intense detail! *Psssshh* — the can opens, the attractive model throws their head back, *Ahhhh*. We share the experience vicariously. Or, we see the burger patty sizzle as the camera pans around it. If we're hungry, maybe our mouth starts watering just a little bit. Again, we're given an experience of the product.

For tech startups, brand is often built at the consumer end through product-led growth — *experience* the product for free yourself, and ideally share what you make with your friends and colleagues (Figma, Notion, etc.). The better the experience and the more people it involves, the stronger your brand growth.

At the enterprise end, though, experience is less direct but no less important. Think conferences (e.g., Dreamforce), charismatic talks (founders on podcasts), learning opportunities (webinars, newsletters), a consistent *and* useful LinkedIn presence (e.g., Gong, as we'll touch on below) — all of these things help add up to reasons why we might actually remember a brand association.

But there's a difference between merely doing an activity and doing it in such a way that creates a meaningful experience for the recipient. Every company posts on LinkedIn; few make it an interesting experience I want to come back for or build into my day. One of the hardest lessons for anyone doing marketing to learn is that just because you're doing the same activities as someone else does not mean you're creating the same experiences.

Think about the personalities and thought leaders you follow on social media, for example. I bet they're great at creating a high-quality, consistent experience you come back for day in, day out. Maybe they give great talks; maybe they make handy infographics; maybe they're just funny people. All of it is about the consistency of experience.

Now, contrast all of that with basic traditional advertising, which involves plastering your logo or product name and an image on static graphics or generic social videos. That's not much of an experience, is it? No experience, no memory, no point. That's why so much advertising simply washes over us, or why, to create any impression at all, it has to be hammered into our minds through sheer repetition.

This is why meaningful, memory-creating marketing is rarely something you can just 'turn on.' If you want to truly build a brand, you've got to have the ability to think through the experience from first to last. You also — more importantly — have to have the budget to give enough people that experience consistently enough for them to both form and maintain a memory association, such that when it comes time to buy, they think of you. That's a rare combination!

Category associations

We don't just experience products and their marketing individually, we experience them as part of a category, too.

Brand memory associations, therefore, also work on a category level. That's why tech companies all tend to adopt similar aesthetics, and why fashion or sports companies likewise all tend to look just like their competitors over time. They're all trying to tap into the memory associations *of the category itself*.

Branding, on a category level, is as much about positioning

your company or product to *fit in* as it is about being unique. That makes sense because it makes it easier for buyers to, well, buy. If someone walks past a shop on the street, for example, it's a service to them that the store looks like the kind of store that it is — fashion shops look like fashion shops; fast food joints look like fast food joints, and so on.

The alternative is aesthetic anarchy, which is a fun idea but not how our brains work. Instead, we rely on visual cues and cultural associations to pattern-match the world around us. That's why B2B SaaS sites all tend to 'look the same,' for example — it helps buyers understand what they're looking at.

Sculpting your brand

Creating unique, memorable experiences while still fitting in with your category can be a fine line to walk. If buyers need common cues to understand what category you're in, how *do* you stand out? How do you become super memorable *despite* those constraints?

There's still enormous room to develop distinctiveness *within* your category, and that's what sales enrichment and automation platform Clay does very well. They have both a distinctive name and visual elements that riff on that name and what it *means*:

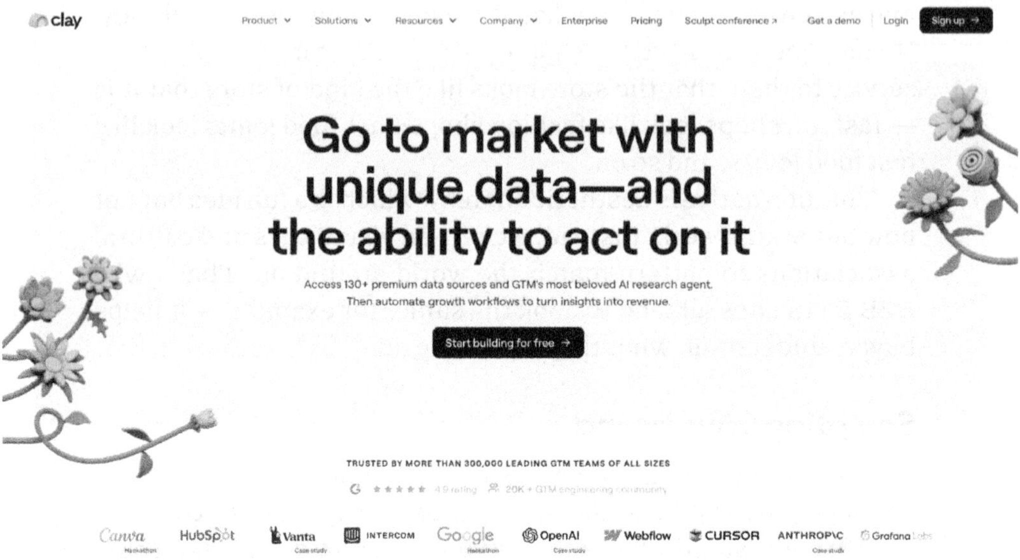

That distinctiveness extends to their conference, Sculpt, where their brand team can really run with the theme, which speaks to the malleability of the tool itself:

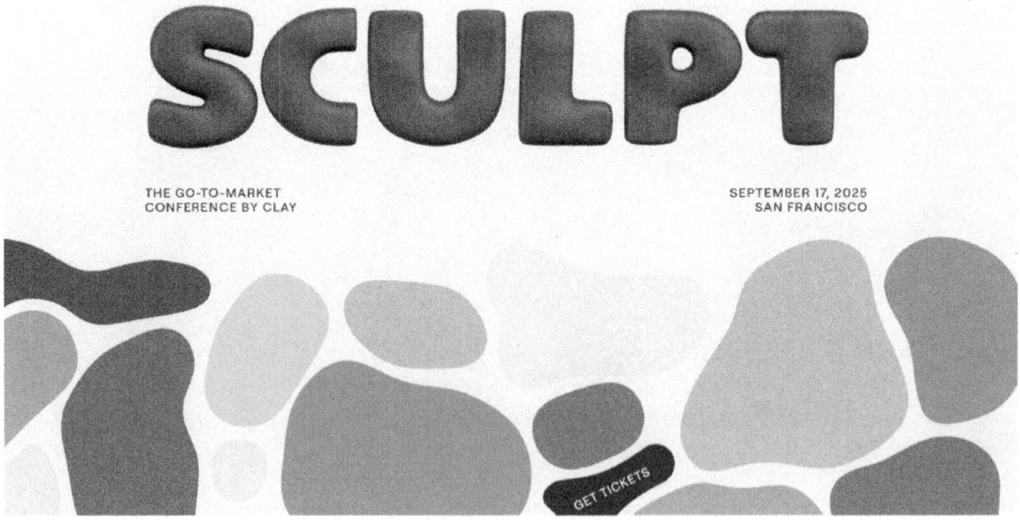

In an interview with *First Round Review,* Clay co-founder Varun Anand says:

> *We've always overinvested in brand way more so than other early-stage B2B software companies. [...] When you look at other early-stage B2B startups, few invest in brand, so everything ends up looking the same. When no one else is investing in something, that's a way to have alpha and stand out in a meaningful way.*

Clay's branding is interesting for two reasons:

1. It's distinct — they have a *look* they own.
2. It runs through everything they do — they have a memorable *position* they own.

On the first point, a branding exercise that results in this kind of (very effective!) work might talk about the brand being *creative, friendly,* and *approachable,* for example. I suppose that's true relative to the very dry and often monochrome world of B2B tech, but the bigger point is simply that it's *distinct*. It works within the norms of the category but creates a coherent theme that's theirs.

That's one step towards being super memorable.

Building memories

Being distinctive matters, but to what end? What's the underlying position you're trying to build in the buyer's mind?

The real meat of branding, including memory associations, is the *need* you can associate with your brand and the scale at which you can establish that association.

Startups tend to think if only their innovation was useful enough, or only the concept they created was relevant enough, then that alone will make them memorable.

That's certainly a part of it, but the more prosaic approach to branding is that it's more akin to *spaced repetition*.

Spaced repetition is where repeated exposure to information builds long-term memories and fights inevitable memory decay. (That decay is predictable by the way — you can model it.)

Spaced repetition is popular with language learning for this reason — see symbol, remember sound. See word, remember meaning. Do that only once, and you'll probably forget it. Do it at set intervals, before the memory decays, and you'll remember it. The same is true for the brand memories we're trying to build.

But for spaced repetition to work, there has to be something consistent, in a positioning sense, that's repeated.

Position-driven branding

It's tempting to think building brand memories -> brand advertising -> running "brand campaigns," which are amorphous things with no real definition.

But if you want to be super memorable, you need to think about branding as compounding activities that work together to build that memory in the mind of the buyer, and that's what we'll explore with Clay in this chapter.

Clay's name is a metaphor (metaphoric thinking is, as we'll see, how we understand the world!) and it comes from its roots as a flexible data manipulation tool, something akin to a spreadsheet connected to various APIs. Hence Clay — a tool for sculpting data.

But that flexibility made positioning a bit of a nightmare.

In a separate (and excellent) *First Round Review* piece by Todd Jackson, Clay's CEO and co-founder, Kareem Amin, says that "getting to the right positioning is often more challenging than getting to the product itself."

That's a message this book can get behind. Quoting Amin (all quotes from *FRR*):

> *This is a common path — a founder starts by thinking a product is cool and then they have to narrow down the customer and the use case, start to tell the story in a cohesive, coherent way, and then the market gets it. The flow isn't always linear; often it's very iterative and messy and that's entirely normal.*

Clay's breadth meant, as Jackson describes, they had too many potential ICPs. But if you read between the lines, you can see the existential struggle they went through around accepting the viable position they'd found in the market vs. their original vision.

The big early right-brain radar vision for Clay was, Jackson writes, "to make the power of programming accessible to more people," and the dream was a broad, horizontal tool that, to my ears, sounds like a Notion or Airtable for API data.

But the vertical they settled on was sales, and the early ICP was *cold email agencies* in, particular. Not exactly a glamorous place to start, but that was the result of their *customer* vision — looking down the left-brain laser beam into who was getting the most repeatable value from the tool.

They didn't own it straight away, however. They didn't turn that insight into a hook. Quoting Amin in *First Round* again (my bold):

> *Maybe three-and-a-half years into building Clay, we were pretty focused on salespeople. But we hadn't committed to that ICP in a way that showed in the product — like how we prioritized new features,* ***or changed the language on the website****.*

The strategic messaging on your website reflects your internal narrative about what you're about and the positioning opportunity you've found, but sometimes it can be hard to swap the original right-brain vision for the more here-and-now left-brain one.

Amin (very admirably) describes the challenge for founders in this situation:

> *In the beginning, I put a lot of pressure on myself with Clay because I wanted it to be both successful and personally fulfilling. As a founder, your company can become your art project, the thing that you want to be actualized through. But this actually decreased the quality of my decision-making because I would get caught in that same spiral of trying to figure out the biggest, most high-impact thing we could possibly do long-term, instead of what is actually useful right now.*

> *Now I've shifted my mindset. I've removed the psychological pressure — it doesn't have to be the biggest thing, it just has to be what other people think is useful.*

That's a very modest assessment of their achievements, given that in 2025 the company was valued at $3.1B. But you can see that battle of internal right-brain/left-brain attention and the struggle for *meaning*. Nevertheless, there's a lot to be said for being super useful for a specific ICP.

You can also see how these big positioning bets rest on a founder/CEO's shoulders — the idea that some lowly product marketer "owns" a company's fundamental positioning in any organization is an attractive fiction that never matches reality. It's a team sport to be sure, but let's be real about who actually calls the positioning shots and how heavy that responsibility can be.

In any case, when it comes to what Clay is fundamentally about, this "own it" lens speaks to something much deeper than brand in a superficial, logos-and-colors sense. Those decisions to trade off the long-term vision for the here-and-now customer vision don't always come easy.

At the end of the day, you need to own a position (and memory) in the mind of the buyer, and those hard decisions ultimately enabled Clay to build the brand, memory associations, and business that they have.

Customer-built growth

That approach to brand is something they've run with, developing a sophisticated ecosystem-driven approach, as Emily Kramer (of the excellent MKT1 newsletter) discussed with Clay's Mishti Sharma & Bruno Estrella.

With a clear ICP, Clay was able to work with freelancers, agencies, and creators to get the word out and help *those* folks look good by showing off what *they* can do with the product.

That created a virtuous cycle where their users *wanted* to be associated with the brand, as that showed how cutting-edge they were, which in turn pulled more people in who could try the product (it's self-serve with sales-assist).

This is reminiscent of the "customer-built growth" concept coined by David Peterson, VC and former growth leader at Airtable, back in 2021. Peterson writes:

> *When end users are empowered to build [through low-code/no-code tools], they also become the first (and most authentic) sales people for your product. They're (understandably) proud of what they built and want to show it off. As they evangelize what they built to their colleagues, they're inadvertently selling your product as well.*
>
> *This dynamic is the essence of customer-built growth. What if every end user could be a node for future growth, not because they want to earn a referral credit by sharing a sign up link, but because they want to show off what they built to their colleagues or peers? Or make a living selling something they built with your product?*

Along with Airtable, Peterson highlights Notion, Zapier, and Figma as being emblematic of this "customer-built growth" phenomenon. Having your users constantly showing off your product to their peers is one way to be super memorable.

While Clay presumably aspired to be mentioned in the same breath as those companies, they found product-market fit in sales enrichment, and in doing so, found something else they could own.

Back to narrative

In serving their customers, Clay has developed what they call a GTM engineering role ("basically a mix of a sales engineer, an AE, and an SDR"), built a narrative around that ("The rise of the GTM engineer"), and are now trying to own their association with this role.

That is, by zeroing in on a niche, they were able to:

- Discover something novel through their own unconventional

approach to solving their needs first (how to demo Clay).
- Give that insight a name (the "GTM engineer").
- Put that owned concept back on the radar of the market more broadly.

In their announcement of a $100M raise, for example, they led with "Clay raised $100M at a $3.1B valuation to power today's breakout AI-native job: GTM engineering."

Yep, they developed their own concept. In our *concept = wave + playbook + tool* formula, their GTM engineering concept is: No-code/AI as the wave, GTM engineering as their owned playbook/role, and Clay as the tool.

The strongest case for this concept is that it's similar to Gainsight owning their relationship with the then-emergent role of Customer Success. The case against is that it's a fake category creation play. In Clay's defence, it's based on what they've seen in their own GTM work; whether the market accepts it, though, remains to be seen.

Either way, Clay seems to be well on their way to building a super position:

- They caught the tail end of the no-code/'customer-built growth' wave and are now riding the AI wave.
- They've found where their unique value resonates (after *years* of zero-revenue exploration).
- They're owning their distinctive brand association with sales enrichment and their emerging GTM engineering playbook.

Narrative pushback

"GTM engineering" has faced some pushback, though. To own an association in the market, the market has to accept that your narrative is based in reality. One difference between today and earlier attempts to coin concepts and define roles, however, is that marketers who are sick of having dubious new concepts put on their radar (and LinkedIn feed) now call them out.

There's an emerging audience for the LinkedIn hater — viral, ChatGPT-assisted takedowns of whatever the trend *du jour* is. Even if the content of these takedowns is still incredibly corny, AI-generated "That's not X. It's Y." prose, they still get likes because the core message of here-we-go-again resonates.

Personally, AI slop aside, I think criticism is healthy. This book includes a takedown of category creation after all. It just means the onus is on the concept creator to not only coin a term but prove the actual value and demonstrate that what they see is real, and not just made up.

That's the challenge for Clay, and that's been a theme of this book and the **ride it** strategy in particular — you *have* to prove your value. You have to own something real, and an industry having a high bar for that proof of that reality is a good thing.

Nevertheless, Clay have still managed to reach the final super positioning stage of combining it, which we'll explore in the **next chapter**. It's not about finding a niche *or* building a brand *or* creating a ~~category~~ concept, it's about learning from the market as you go and making compounding decisions to build something that's greater than the sum of its parts.

B2B 'physical' availability

With the right mix of ICP focus and a wave to ride, Clay eventually found a brand association — or series of associations — they and their ecosystem partners could build, ultimately creating a high degree of 'mental availability' in the market. When marketers and sellers thought of enrichment and AI tools, they thought of Clay. Even the mild pushback is a sign of high mental availability — when your content is worth commenting on, you've reached a place of significance in the market's collective mind.

In the B2C world, however, mental and *physical* availability have to work together. It's one thing to remind folks that a sweet, sugary, caffeinated cola exists with a certain logo through sustained advertising (mental availability), but you also have to make sure it's physically available when the need strikes (vending machines, supermarkets, fast food partnerships, and so on).

What does that look like in B2B? We touched on Clay's domination of the social environment, but what about search?

For the B2C and SMB-focused design platform/juggernaut that is Canva, that meant dominating SEO in the early days. It meant showing up for every Google search around a design job-to-be-done, ever. That was physical availability for "How to design a [card/resume/poster]" in the digital world.

These days, the race is on to get recognized in AI search (i.e., ChatGPT) for much the same reason. No matter how carefully you've thought through your positioning, you still need to be 'physically available' when it counts.

In sales-led B2B, that can take many forms, but the simplest way to frame it is: are you easy to recall and easy to buy? That means *eventually* being there in webinars, on LinkedIn, in the search engine result pages, at the trade shows, and so on, perhaps with a targeted outbound motion that means good-fit prospects actually encounter you.

With all of this surface area to cover, you need more than just your latest feature release or case study to talk about, too. That's why having a strong *concept* to work with is so important. And that's why Clay will always get more mileage out of "GTM engineering" for brand building and owning a message than whatever integration or feature they launch in a given month, however useful they may be. Having a concept that's bigger than your product just creates more *meaning* for your brand and gives you more to talk about, all of which keeps building and refreshing those memory structures.

Brand > advertising

"Owning it" is ultimately about branding beyond mere advertising. The B2C folks will be the first to admit that advertising is a weak force, especially in the short term, so think long and hard anytime someone suggests you need to run a one-and-done "brand" campaign.

Running sustained brand campaigns can be worthwhile, though, just remember that the very best brand ads also tend to activate buyers. That is, it prompts them to buy in the short term, not just remember your product or brand over the longer term, which means they can be run indefinitely. (That's why Squarespace was so successful with their podcast 'brand' ads, as we'll touch on below.)

Advertising might be the only real lever B2C brand advertisers have to pull, but in tech, especially startups, we can build the product, playbook, UX, and entire buying experience to match the memory we're trying to build.

That's something Linear founder Karri Saarinen knows very well, which he explained on Twitter/X:

> *I was reflecting back to my time at Airbnb. I remember for [Airbnb founder Brian Chesky], the brand was the most important thing and I think it's still very rare that tech company founders understand what is brand and how to use it.*
>
> *I also think the brand is as much internal as it is [an] external thing. The brand is essentially what you are as a company, and internally it affects the culture and the outputs.*
>
> *Many tech companies push products and messaging which doesn't seem to make any sense. When Apple puts out something, it 99% of the time feels like Apple. It's not by mistake.*
>
> *The consistency comes from the fact that everyone in the company [has a] very clear understanding [of] what the company is, what it stands for, what's the quality bar, what it does, how and why. Consistency is important for a brand, it's like the soul or personality of your company. [...]*
>
> *At [L]inear [we] started with the brand from day one.*

It's that *consistency* that creates memories, and that consistency isn't just in a logo or ad; it's in all the marketing output, it's in the product, it's in the culture, and it all amounts to a more implicit, right-brain *gestalt* that makes you a coherent entity in the prospect's mind.

That's what owning it means when it comes to brand as a positioning strategy — creating a distinct memory association over time through impeccable standards and consistency.

Memory decay

When it comes to brand, failure state for most sales-focused tech outfits is ignoring brand entirely, thinking they're too specialized, too sales-driven, too industry-specific for anyone to care about their 'brand.' That's true insofar as you don't have to be a *cool* brand, but ignoring brand altogether is a self-limiting belief.

This becomes a particular problem when executives try and cut costs, so they cut back on marketing activities. Initially, nothing changes. *'Aha! I knew all that marketing stuff was fake!'* they think. (Sometimes it is!) But often, there's no change because the business is still profiting off brand memories built and sustained over *months and years* prior.

Eventually, however, those memories decay. A couple of quarters later, pipeline coverage doesn't look so good. Marketing is ordered to *'Go out and get some leads!'* But if there was a magical 'Generate leads' button, they would have pressed it already. Spinning up marketing activities to build memories takes time. And it takes longer still for buyers to then become ready to buy *and* go through your months-long sales process.

Advertising to the 95%

This brings us to the EBI's 95:5 rule of brand marketing by the way, which is a LinkedIn favorite.

The idea is that, as a loose rule of thumb, only 5% of a given B2B market is actually 'in market,' i.e. looking to buy, at any given time. And when they do come in market, most folks search their memories first.

Marketing is, then, largely aimed at the other 95% to build and, crucially, *sustain* memories so when they are ready to buy, they think of you.

The 95:5 playbook

What does this look like? Devin Reed, a content marketer for Gong, detailed their playbook on HubSpot's blog and summerized it on LinkedIn. Reed claimed that:

> *I spent a year at Gong creating content for the 95% of*

buyers who weren't ready to buy. We grew from 12K to 220K LinkedIn followers and generated millions in pipeline — and never talked about our product.

They may have never talked about their product, but they absolutely talked about the things they wanted you to *associate* with their product, like how you do sales effectively.

Also, note that this is marketing and sales folks talking about selling to marketing and sales folks so they can... sell to marketing and sales folks (LinkedIn gonna LinkedIn), but you can see the principle in action.

In a sense, social — and LinkedIn specifically — is the new inbound and content marketing strategy for this market.

Marketers once spent years creating content hoping to rank for keywords; now they spend years pumping out content through LinkedIn (and newsletters, podcasts, and so on) so they can build that name/need memory association in your mind. Need sales tools? Think Gong.

You search your memory before you search Google after all, and when you do search Google, chances are it's then a brand search.

Associated with what?

If B2B positioning — and building a super position — was merely a matter of running the right brand ads to build your brand association, it would be a much easier game.

The reality is that it's a compounding game that effects every part of your operation.

That's why there's a lot to be said about caring about consistency — in product, your message, your buyer and customer experience, and ultimately the name/need memory association you create is low-hanging fruit that benefits the whole business. Just care more.

That way, if nothing else, folks who may have interacted with you but didn't buy from you this cycle may well come back next time they're in-market, when they switch jobs, or when they want to make a recommendation to a friend, simply because they felt positively about you and remembered you because your brand wasn't the equivalent of a manilla envelope stapled to a beige wall.

Owning it from the front

The thing is, though, that leadership has to own it. They're the gatekeepers of consistency, after all. This consistency of experience doesn't happen on its own, especially if they're treating Marketing like a leads piñata.

The bar needs to be set by leadership so everyone knows what standard they're aiming for, what message they need to stick to, and ultimately what kind of experience is acceptable to deliver. (That's why a simple strategic narrative can go a long way in driving alignment, too!)

Memory builders — examples

Brand building, then, is essentially putting something on people's passive right-brain radar consistently and over a long enough time period that a clear name/need association is built — a memory that sticks in the buyer's mind.

Incumbents have it easy — by virtue of being the biggest, they become the safe option that 'everyone knows', and continue to build and maintain memory associations in the market simply by continuing to remind people they exist.

Nevertheless, from a startup point of view, how do you use this kind of memory-building as a growth tool?Here are a few possibilities.

Squarespace — exploit a new channel

Squarespace built one of the strongest brands in SaaS over a number of years through their innovative, relentless, and exhaustive use of podcast advertising.

By seemingly advertising, at one point, on every podcast, ever, they built incredible brand memory associations in the market. Want a website? Have a friend who does? Chances are you think of, or recommend, Squarespace. Why? Because they're first to mind.

This is the almost Pavlovian response that effective brand advertising over time generates. And in Squarespace's case, it also generated immediate conversions, so they could keep doubling down on it. Their ads performed 'double duty' to use marketing-speak, building memories for the future *and* converting customers in the present.

Slack & Linear — double-down on quality and innovation

Past behavior and experience drive attitudes towards a brand, not the other way around, much to the chagrin of marketers who think they can imbue their brand with magic powers of persuasion.

Think about ChatGPT, for example. Super weird name. Standard logo. Generational innovation anyone can use for free. You can guess which of those drives attitudes for the brand.

Strong brands in tech are also often built on a superior user experience, and that was true for both Slack and Linear for that matter. Both products were riffs on existing categories (team chat and issue management, respectively). But both founders had an acute sense of their *experience* mattering, and that set them apart.

For Slack, the operating metaphor was that of impeccable restaurant-like service with a high bar for quality, which was established by founder Stewart Butterfield in a classic memo sent out before the product launched (as we touched on in **chapter 10**).

Likewise, for Linear, as we touched on above, the brand was every interaction folks had with the product and company, including a stunning homepage and incredible product UX.

The key here is the contrast — the common alternative to Linear was the long-in-the-tooth Jira, and Linear was positioned as a breath of fresh air for serious professionals who care about their tools.

This made them the first choice for early adopters who wanted to use the hottest tools. That got their momentum going, and that established their early position as the new default choice for discerning product development teams starting new projects.

Notion — go B2C2B

Brand tends to work more at the B2C end of the spectrum, and some of the best SaaS brand plays, therefore, tend to look very

B2C-ish.

Take Notion, for example.

We discussed Notion as "sugar-coated broccoli" in the previous chapter, because Notion is, on paper, a pretty weird concept — docs and databases intertwined in an innovative canvas of sorts where users can create their own simple tools and workspaces.

What's interesting, though, is their B2C2B approach. They go *broad* with incredibly accessible branding and marketing, get consumers hooked on the tool, grow their community, and then hope they'll bring the tool into their workplace, which is when Notion cashes in.

That only works if the brand sticks, though. There needs to be that strong need/name association, and Notion has built just that for their project, work management, and now AI tools.

Brand as a forcing function

Brand can be a great forcing function in terms of making you think hard about exactly what clarity you want to create in the market — what need association *do* you want to establish in prospects' minds? What do you want to be famous for? What's your pithy from/to transformation or your hook?

Speaking of hooks, this is why value blobs don't work — they don't create a clear name/need association. What's one term you can be associated with? (For me, I hope it's 'super positioning' as a differentiator and positioning as a category association — we will see!)

If that association is going to take years to build, once you know what it is, the sooner you can get started, the better.

Next, we'll look at what happens when everything comes together in a winning super position.

What's your memory association?

- What's your vision for the distinct value, or need, you want prospects to associate with your brand?
- When people think about your brand now — *if* they think about your brand — what's the current associated need?

- If you surveyed 10 or 100 customers, what do you think the consensus on the associated need or attribute would be? Would there be a consensus at all?
- Is there a channel or unique GTM opportunity in your market where you can show up consistently to drive home a unique name/need association?
- How relevant is brand to how your prospects buy? Do they buy by searching their memories first, and if they do, what do they say triggered that mental search? How can you strengthen your association with that trigger?

14

Combine it & build a super position

Combine it in a nutshell:

- **Goal:** Find a way to combine being **relevant, useful, and memorable** so you are **known, chosen, and remembered**.
- **Strategy:** Once you've found a winning super position, ride it to be #1 in your niche, then your market, then your adjacencies.
- **Attention:** A combination of right and left-brain attention, with brand building to boot.
- **Action:** Combine it — keep your eye on the prize and sequence your positioning as you go from niche to reach.
- **Position:** A compounding stack of *riding it, finding it,* and *owning it.*

The last of our four super positioning strategies is, essentially, "hold on tight." If you've found a winning super position, where some or all of our core positioning ingredients come together, there's still a mountain to climb, but the good news is you *have* the mountain to climb.

Lots of folks, however, get caught up on the next deal, the next release, and the next funding round without stopping to think about the hill they're trying to climb, what truly winning their segment or their market would plausibly look like, and how they can position themselves to get there.

To be fair, some folks don't have to. If you've managed to tap some rich vein of demand and you're being inundated with customers, there's no need to worry about building momentum through niche positioning — the momentum is already there. You might think about what comes next, of course, and be deliberate in how you position in new adjacent niches as you grow — there's always the next level to reach and the next boss to beat — but when you're on this trajectory, it's really just up to you to not snatch defeat from the jaws of victory.

But if you haven't stumbled across that pot of demand gold, it can be a highly useful exercise to think through what, exactly, your vision is for the winning super position you'll build for your company, with your product, in your market.

What does your winning super position look like?

Consider making a checklist of steps that would help you identify when you're on the right track, even if it's just the basics, for example:

- You've clearly identified the goals you need to achieve relative to your ambition and funding.
- You've planned a reasonable positioning experiment to create the necessary momentum to reach those goals.
- You're getting an appropriately positive reaction in demos and sales calls — people are genuinely lighting up and wanting more and not just giving you the polite "Oh wow, sounds great,

yeah, uhh, we'll definitely get back to you on that, just let me have a think about it" (i.e., no, thank you).

From there, it's a matter of what makes sense for your GTM motion.

Sales folks can 'segment & sell' with a narrow focus in a niche to build momentum while product-led growth practitioners may go for 'cast & catch' with a broader pitch for an entire discipline (like product design, marketing, or general productivity and team collaboration) in a much more horizontal fashion.

Either way, your positioning, ambition, and go-to-market strategy all need to work hand-in-hand.

Figma's super position

A super position is where a strong change narrative meets a laser focus on the actual tool, while building an incredible brand in the process.

On the PLG end, Figma, for example, did this extraordinarily well. They managed to:

- **Ride it:** They managed to be super relevant by riding the rise of digital product teams, design systems, and the "Design Systems Manager," studying how modern product teams work so they could prove their value.
- **Find it:** They managed to be super useful by having an extreme laser focus on their product and technology, zeroing in on the appropriate use case for their WebGL-powered tech, which competitors simply couldn't copy.
- **Own it:** They managed to be super memorable thanks to the collaborative, multiplayer nature of the tool that often drew in the entire org (network effects ahoy), exposing everyone to the brand and Figma experience.
- **Combine it:** They found a winning super position, one that ultimately had Adobe licking their lips in anticipation of acquiring them for $20B, which didn't work out, so they went public instead, and now have a market cap of ~$26B.

Figma's complete dominance of their category was something

to behold, perhaps as best summed up in the chart below from UX tools.

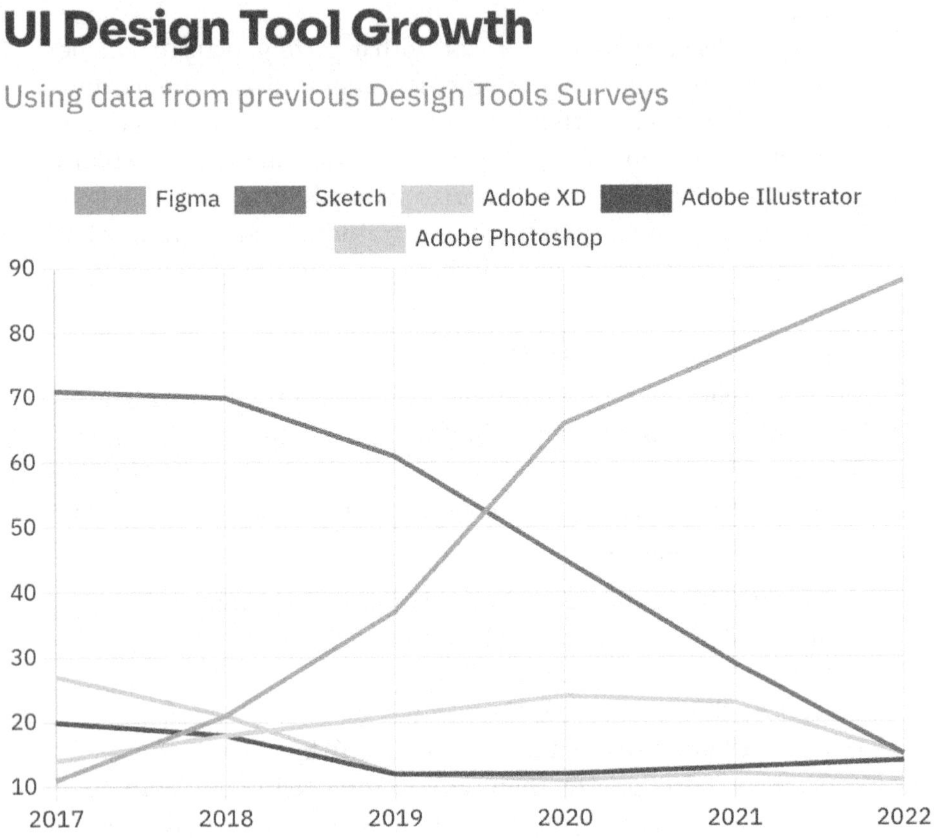

And to be fair to Figma, they did a lot of analysis and strategy work to pull that off. In an interview with Connie Yang on the *Distributed Podcast*, Figma CEO Dylan Field said (edited for clarity):

> *There was a point where we had to say we're going to focus on interface design [...] We used a cool framework — omit, raise, reduce, create — from* Blue Ocean Strategy *to figure out, okay, what are we doing that's different than the*

competition? What are we not doing that the competition is doing? What are we doing worse and where are we going to make this brand new?

Props to their use of the *Blue Ocean Strategy* framework there!

Qualtrics' winning position

On the sales-led and enterprise side, it's worth returning to Qualtrics, which we've touched on a few times because they're also a fantastic example of making these strategic positioning choices over the lifetime of the company, which was founded back in 2002. Specifically:

- **Find it:** Qualtrics started as a useful survey product, originally selling to *marketing professors in business schools* because that was the background of one of the founders.
- **Ride it:** 15 years (!) into their journey, Qualtrics managed to generate insane relevance for a survey platform through their concept/category — "experience management." The brief wasn't to create a literal new category of survey software; it was to transcend the survey software category entirely with a new concept they could put on the right-brain radar of C-suite prospects. The wave they identified was the "experience economy," and they aimed to prove their value with their experience management, or "XM" playbook. (And they could prove it because they had 15 years of experience under their belt.)
- **Own it:** They went all-in on "XM" as a brand (going so far as to make XM their stock ticker, echoing Salesforce's CRM ticker). XM is in their logo; it's how they organize their product lines ("XM for customer experience," "XM for employee experience," etc.); it's the theme for their events and content marketing, it's everywhere. That's what it takes to make your brand super memorable (at least in the world of enterprise SaaS!).

All told, it's an incredibly strong super position that hammers home their owned message and position across every touch point.

But it took years to build, and it was initially a lonely time trying to make it happen.

This took Qualtrics to an IPO and subsequent acquisitions — acquisitions that were predicated, in part, on the strength of their 'category creation' efforts.

It also drove their main competitor, SurveyMonkey, bananas. They tried to rebrand as "Momentive" to chase the enterprise and the "experience management" concept Qualtrics pioneered, only to revert back to SurveyMonkey two years later.

Evolving your brand to set the pace and completely rebranding to chase a leader are two very different strategies with very different results. High-stakes rebranding is a fraught business — it's not a positioning experiment I'd recommend.

In this case, Qualtrics evolved their brand to build incredibly tight memory associations between their name and XM.

SurveyMonkey's attempt to try and build a new brand from scratch that could also be associated with experience management was, in comparison, always a doomed mission. Especially because "experience management" was never a real generic category of software they could compete in — it was always an extension of the Qualtrics brand. SurveyMonkey could never 'own it' like they did.

Winning positions are unique

Qualtrics is an interesting example of the fundamentals of super positioning, but I'm not describing it here as a template every enterprise-focused B2B SaaS company must follow.

Instead, I include it to highlight how many things need to go right.

And that's going to be true for you, too. Winning positions are ultimately unique because waves of technology (i.e. what you can ride) and emergent opportunity (i.e. what you can find) are unique.

Clay, as we saw in the last chapter, is on the way to building a super position precisely because they saw things that were unique to them — the outbound enrichment opportunity, the wave of AI research agents, the "GTM engineering" concept, and, of course, their data-sculpting spreadsheet-and-APIs product.

An *insane* number of things have to go right as you ride a wave, find a niche, own a memory, and combine it into a super position, but the result will be unique to you — all great business positions are.

Great businesses 'compete to be unique,' to borrow the line from Michael Porter that sums up just about every popular business book that has been released since. It's not easy — it's brutally hard! — but we're insanely lucky to be able to participate in this crazy world of tech and startups at all.

Complexity from simplicity

In the world of tech and venture-backed startups, this is what we're trying to do — we're trying to build truly great, insanely unique businesses. That requires a great deal of vision from you as a marketer, GTM leader, or founder, both into what is and what could be.

But these big winners aren't born with this level of strategic complexity, however. Instead, by necessity they start from something small that works, and while I'm throwing ideas at you, dear reader, I'm really just hoping there's one idea that clicks.

Again, the point is to find a simple system that works to build on, per Gall's law:

> *A complex system that works is invariably found to have evolved from a simple system that worked. A complex system designed from scratch never works and cannot be patched up to make it work. You have to start over with a working simple system.*

When discussing successful businesses, there's always the risk of looking at famous companies and repeating the "How to draw an owl" meme, where there's mountains of context missing between where they started (a couple of rough circles) and where they finished (a beautifully detailed hand-drawn owl).

But in the spirit of keeping things simple, let's think back to where we started, with hooks and catchy ideas.

Hook as super position

These were the hooks we ran through in **chapter 3**, for example:

- Use cloud software not on-prem. *(Salesforce's "End of Software" campaign at the turn of the millennium, before either 'the cloud' or 'SaaS' had been coined.)*
- Use chat not email. *(Slack.)*
- Use chat not forms. *(Drift and "conversational marketing.")*
- Do 'inbound' marketing. *(HubSpot, riding the digital marketing wave.)*
- Use agile not waterfall. *(Jira riding the agile wave.)*
- Use Linear not Jira. *(Linear riding a classic anti-incumbent wave.)*
- Mine sales calls for insights. *(Gong, riding the call recording wave.)*
- Design collaboratively in the browser. *(Figma.)*
- Build your own workspace. *(Notion.)*
- Payments for developers. *(Stripe.)*
- Code with AI. (*Cursor et al.*)

But these weren't just hooks.

Ultimately, they turned out to be super positions.

And while the products and businesses built *around* these positions are very complex, the core idea — and the position built in the mind of the buyer, either explicitly or implicitly — is often very simple.

One way to keep it simple is to come back to this basic question: How are you helping your customers succeed?

Answering that question — articulating your theory of customer success and finding your hook — can help bring you back to that simple system that works that a complex system can evolve from.

Down to you

It requires someone to do the looking, so to speak. And now we know how the underlying fundamentals of attention work (including your vision) and how they manifest in B2B positioning, I hope you can be the one that sees that new theory of success, that hook, and, ultimately, that super position.

I mean, if not you, who else? At the end of the day, it's up to you and the vision of your company to look *out* on the right-brain radar and *down* with your left-brain laser beam, and bring that all together in a memorable brand association. That's the super position we're shooting for.

That's always difficult, always fraught, and always involves scratching around to get started. But knowing how positioning fundamentally works — along with the hook in the buyer's mind you're gunning for — at least, I hope, helps you know *what* to look for.

Likewise, I hope that, in some small way, the four positioning strategies we've discussed — ride it, find it, own it, and in this chapter, combine it — have also given you a better sense of what it is you're looking for so you too can build a truly unique product, company, and position in the mind of the buyer.

Let's wrap this up with a few final questions to ponder and then I'll wrap up the book with the super positioning manifesto next.

Creating your super position

- What's your current vision for the winning super position you want to achieve in your market?
- What choices will you make to get there, especially in terms of proving value on a wave, finding it in a niche, or owning it as a brand?
- How can you stack your advantages across the wave you're riding, the technology you're developing, and the brand you're building, like Figma did, for example?
- Or, how can you sequence your growth, going from niche to reach to brand, like Qualtrics did?
- And ultimately, what's unique to you and the story you can tell, as you compete to be unique?

PART V

MANIFESTO

15

Super positioning manifesto

I hope you've learned some useful ideas from this book. The process of writing *Super Positioning* (and *Positioning Playbook* before it) was, as you can imagine, a period of intense learning for me, too.

Let's finish where we started — on the importance of getting specific. Here are a bunch of specific beliefs I landed on that make up the super positioning manifesto.

I know we're not trying to overthrow regimes here — just sell some software — but consider this a snapshot of the themes that drive *Super Positioning* that you can come back to when you need to refresh your memory.

Here we go.

Positioning is a system

Systems > spreadsheets > statements. Positioning isn't a statement (or spreadsheet) you wish were true; it's a system that works across your GTM motion for *making* a position in the mind of the buyer come true. Systems over statements.

Two modes of attention

The crux of positioning as a system — and the most important concept in this book — is the two modes of attention we all use. Two modes of attention mean:

- **Marketers:** Two distinct approaches to positioning (narrative vs. competitive positioning, or story 1 vs. story 2).
- **Sellers:** Two distinct ways customers show up in calls (radar vs. laser beam).
- **CEOs/founders:** Two distinct ways for you (and your investors) to see your company (outward vs. downward vision).

It affects everyone and everything. Especially in GTM.

Known, chosen, and remembered

By using and engaging those two modes of attention and creating memory associations, we want to build a narrative (the thing you're known for), ensure we're chosen (we're a better choice relative to competitors), and remembered (customers form a brand memory and share that with others).

This is not just the job of marketing — though it's what marketing exists to do — it's the job of the entire organization, *especially* if you're a startup.

Startups are about winning

The point of achieving those jobs of being known, chosen, and remembered is to **win**. There are multiple ways to win, but if you're a venture-backed startup, everything must be seen through the lens of winning. Startups are about winning (or dying trying), and winning doesn't come through headline tweaks or shallow positioning statements, it comes from **executing on strong ideas, clearly expressed.**

Startups are about compounding

Super positioning is the antidote to silver bullet thinking that often pervades marketing and startups. Success doesn't come from One Weird Positioning Trick.™ It comes from finding a viable position as early as possible, creating a solid foundation. Venture capital will, as Eric Paley described, compound the strengths *and* weaknesses of your position, so better to make it as strong as possible as soon as possible.

Product precedes positioning

Product is the most important thing. Go-to-market starts with a product and ends with a position — or memory — in the buyer's mind. The question is, is that a favourable memory? As Bill Bernbach said, "Nothing kills a bad product faster than good advertising."

The point of the product is the innovation

Invention becomes innovation when it's well-positioned. Inventing some new technology or product is great, but that invention becomes a value-creating innovation **when it's positioned appropriately in the market.** In B2B, your innovation then drives productivity, and productivity ultimately drives prosperity and increases standards of living — a good thing!

Strong positions are built on unique concepts

Your company needs to compete to be unique across as many dimensions as possible. You need a product that's different from the competition. You need a GTM motion that's different from your competitors. You need a concept that's bigger than your product. You need to find the simplest system that works to start (Gall's Law), and then start stacking compounding advantages.

Better outcomes don't require more money, *they require better ideas*

Poor returns come from poor ideas. We've all seen what a decade of so-called "category creation" combined with near-infinite VC dollars has produced — massive waste and foie gras'd startups. New ideas — and better returns — come from seeing the world in a new way. Understanding how vision fundamentally works, i.e., how we attend to the world around us, unlocks that ability to see new opportunities.

If you want to create a position in the mind, *you need to understand the mind*

This means rejecting neurobabble while embracing the profound new neuroscience on how we fundamentally attend to the world as humans — with two brain hemispheres that **produce two fundamentally different modes of attention**.

Efficient GTM motions *address the mind as it is*

Two modes of attention mean two fundamentally different types of stories exist. We need both — the big change story *and* the competitive edge, all wrapped up in a memetic hook that gets remembered. Your GTM motion needs to be built to serve those fundamentals.

Super positioning starts with a narrative

People only care about why you're different *after* you've appeared on their radar in the first place. So that's where super positioning starts — with a narrative — and ends in a hook.

Super positioning > narrative-ONLY positioning

Big stories are just tall tales if they don't translate to actual differentiated value.

Super positioning > niche-ONLY positioning

If positioning is only about niches and never about narrative, then you'd never get HubSpot, because they would have given up after starting life targeting the legal niche as LegalSpot.

Super positioning > fractured positioning

Our brains process information from the whole and what's new (narrative), to the parts and pieces (differentiation), and back into an integrated memory (hook) — or it's discarded. This is the R/L/R jump. Modern positioning treats each as a separate discipline (CEOs and sales doing narrative, PMMs doing competitive positioning, and brand marketers doing brand). In super positioning, **all three prongs must work together.**

Super positioning > branding

Brand is about a memory association in the mind of the buyer. That's the ultimate destination — having people buy you simply because they know you, are happy to choose you, and remember you when it comes time to buy. Visuals, colors, and abstract shapes (i.e., logos) help create connections, but they don't convey much specific meaning. Your product and your words, on the other hand, do.

Super positioning is a unifying theory

The super positioning system resolves the tension between the competing positioning schools of thought (go big — category creation; go narrow — niche down; go broad — maximize awareness). This gives us a unified theory of positioning as a concept and should shape how we fundamentally see opportunity and build startups.

Super positioning is a creative act

Most positioning gets stuck in competitive analysis, when the real job of positioning is stacking unique creative choices that make you different. Analysis matters, but it can only analyze what you (and your customers) have already created, and that underlying act of creation is what startups, products, and ultimately positioning, is all about.

I hope these beliefs — and this book — have resonated with you!

Thank you so much for reading!

I'm honored you picked up *Super Positioning* and I truly appreciate you making it to the end. I'd love to hear what you thought of the book. Don't be a stranger, drop me a note at: **luke@superpositioning.co**

If you want to hear more from me on this topic, check out my earlier textbook-length deep dive, *Positioning Playbook,* on Amazon.

And if you need help with any of this stuff... that's what I do! Drop me a line and let's chat:

- My services: **https://superpositioning.co**
- Email me: **luke@superpositioning.co**
- Connect on LinkedIn: https://www.linkedin.com/in/ldstevens/

Thanks again!
— Luke

Made in the USA
Monee, IL
04 November 2026

67379707R00144